# Common Rangeland Plants of West Central Texas

# Common Rangeland Plants

George Clendenin

of West Central Texas

USDA–Natural Resources Conservation Service

TEXAS A&M UNIVERSITY PRESS
College Station

Editing, design, and composition copyright © 2016 by
Texas A&M University Press
All rights reserved
First edition

This paper meets the requirements of
ANSI/NISO Z39.48-1992 (Permanence of Paper).
Binding materials have been chosen for durability.

Manufactured in China by Everbest Printing Co.
through FCI Print Group

∞ ♻

Library of Congress Cataloging-in-Publication Data
Names: Clendenin, George, author.
Title: Common rangeland plants of west central Texas / George Clendenin.
Description: First edition. | College Station : Texas A&M University Press, 2016. | Includes bibliographical references and index.
Identifiers: LCCN 2016016795 (print) | LCCN 2016020743 (ebook) |
ISBN 9781623493912 (flexbound : alk. paper) | ISBN 9781623493929
Subjects: LCSH: Range plants—Texas—Concho River Watershed—Identification. | Range management—Texas—Concho River Watershed. | Wildlife management—Texas—Concho River Watershed.
Classification: LCC QK188 .C57 2016 (print) | LCC QK188 (ebook) |
DDC 582.1309764—dc23
LC record available at https://lccn.loc.gov/2016016795

DEDICATED TO LOUISE JONES

*To my mother-in-law, Louise, a true lover of bird and bloom, who went to the Lord while I was finishing this book. Thank you for passing this affection on to your daughter, my wife, Diana.*

Cover photo: Livestock ranch in Water Valley, Texas. Photo by George Clendenin, USDA-NRCS.
Dedication: Wildflowers in bloom near San Angelo.
Table of Contents: Sideoats grama, the State Grass of Texas. Photo by Dee Ann Littlefield, USDA-NRCS.
Acknowledgments: Mealycup sage. Photo by Dee Ann Littlefield, USDA-NRCS.

Publication of this book was made possible by:

*Concho Valley Resource Conservation & Development Council*
*Grazing Lands Conservation Initiative in Texas*
*USDA–Natural Resources Conservation Service*
*Tom Green Soil and Water Conservation District*

# Contents

Acknowledgments / xi

## 1 Overview of the Book
Featured Plants / 3
Geographic Area Covered / 4
Climatic Data / 4
Major Land Resource Areas of the Concho Valley / 8
Photographs / 12
Plant Names / 12
Toxic/Poisonous Plants / 13
Noxious and Invasive Plants / 14
Ethnobotany of Plants / 17
References and Literature Cited / 18

### SECTION I

## 2 Common Rangeland Plant List
Grasses and Grasslike Plants / 23
Forbs / 24
Shrubs and Trees / 27
Vines / 28
Cacti and Agaves / 29

## 3 Plant Profiles
Grasses and Grasslike Plants / 33
Forbs / 117
Shrubs and Trees / 339
Vines / 431
Cacti and Agaves / 457

## 4 Comparing and Contrasting Similar Plants
Comparing Western Ragweed (*Ambrosia psilostachya*)
   and Field Ragweed (*Ambrosia confertiflora*) / 477
Comparing Bush Sunflower (*Simsia calva*) and
   Orange Zexmenia (*Wedelia texana*) / 478
Comparing the Catclaws: *Acacia* versus *Mimosa* / 479
Comparing Western Soapberry (*Sapindus*), Pecan (*Carya*),
   Little Walnut (*Juglans*), and Flameleaf Sumac (*Rhus*) / 480

## SECTION II

**5 Rangeland Management in West Central Texas**
 Historical Accounts of Vegetation and Landscape, 1683 to 1858 / 485
 Overview of Livestock Grazing Management / 494
 Reading the Landscape in West Central Texas / 497
 Managing Riparian Areas in West Central Texas, by Steve Nelle and George A. Clendenin / 504
 The History and Use of Fire in Texas and the Edwards Plateau, by Butch Taylor / 513
 Prescribed Burning in West Central Texas / 527

**6 Wildlife Management in West Central Texas** by Steve Nelle
 Managing for White-Tailed Deer in West Central Texas / 537
 Managing for Turkeys in West Central Texas / 544
 Managing for Doves in West Central Texas / 548
 Managing for Quail in West Central Texas / 550
 Birds of West Central Texas and Their Management / 554

**Appendixes**
 APPENDIX A: Leaf Shapes and Types / 565
 APPENDIX B: Tank Mixing Guide for Chemical Applications / 566
 APPENDIX C: Common Conversions / 567
 APPENDIX D: Livestock Husbandry / 568

Glossary / 569
References / 573
Index of Plants (Common and Scientific Names) / 577

Stewardship—\'stü-ard-ˌship\: the conducting, supervising, or managing of something; especially : the careful and responsible management of something entrusted to one's care, as in stewardship of natural resources (from *Merriam-Webster* online dictionary).

# Acknowledgments

From this book's inception, it has always been difficult to describe the vision, the purpose, and the content. These are crucial moments for accepting or rejecting a vision. Sometimes it takes just one other person to provide that needed affirmation. There is one person who has stood by me from the beginning: my wife, Diana. I am indebted to her for her generous and consistent encouragement and support through this entire effort, and for her prayers. I also humbly offer thanks to the One who heard our prayers.

I have been honored to serve among the top experts in various fields of ecological sciences in Texas. Steve Nelle, biologist (retired), USDA-NRCS, was there from the beginning to serve as a trusted adviser and mentor during this entire effort. Steve was the senior technical expert for plant identification and management, wildlife management, and riparian science. I salute the lifelong work and passion of the rangers that have gone before me, notably Dr. Jake Landers, whom I consulted early on for plants and content. Thanks to Charles "Butch" Taylor, Regents Fellow and professor, Texas A&M AgriLife Research Station–Sonora, who contributed his expertise on prescribed burning and fire culture in Texas. I owe much of my knowledge and expertise in the field of prescribed burning to Butch. My sincere appreciation goes out to Dr. Terry Maxwell, biologist with Angelo State University, for making his research freely available to me. I am also grateful for the work of Dr. Charles Hart, whose work I consulted for the toxic plant information.

My thanks to the San Angelo field office and area staff of the USDA-NRCS, especially Ronald Crumley and Jaime Tankersley. Thanks to Dee Ann Littlefield for her talents and continuous support. Thanks to Tony Resendez and the Concho Valley Resource Conservation and Development for their enormous financial contribution and support and to the Grazing Lands Coalition Initiative for their additional funding contribution.

A special thanks goes to Shannon M. Davies, director at Texas A&M University Press, for her support and encouragement for this manuscript, and making this dream come true.

And last, the greatest cultural resources in this book are not the plants, but the people. Thanks to the ranchers of the Concho Valley for letting me come through your gates and welcoming me on your lands. You are my friends. I hope this book can provide you with a valuable reference tool and a token of my appreciation.

# Overview of the Book

## Featured Plants

In my capacity as a US Department of Agriculture (USDA) rangeland management specialist, or "ranger," I have had the pleasure of working with many ranchers and of having access to hundreds of thousands of rangeland acres in west central Texas. I have been truly honored by this experience, and many of these ranchers I now call friends. I have traveled deep into these rangelands on foot, on horseback, and by four-wheeler and pickup truck. This book offers an experience-based approach to identifying and managing some of the natural resources of this region.

The Concho Valley is a local name given to an area in the Concho River watershed that encompasses over 9 counties in the heart of west central Texas. The Concho Valley is a ruggedly beautiful and unique territory where west, north, and south Texas seemingly meet on the western porch of the Edwards Plateau. I have chosen this region and this watershed as the basis for this book, although the plants themselves represent a much larger region.

A checklist of the plants of the Concho Valley of west central Texas, originally published in 1998, listed over 980 plant species for this region (Amos 1998). I have highlighted in this book over 212 of the most significant plants in the Concho Valley of west central Texas. The plant list that I offer comes from "boots on the ground" experience of plants that I most often found in my work in this region. What makes a plant "significant" enough to be included here? Plants were selected for this guide if they are native and grow here, if they are native and should grow here, if they grow here but shouldn't be here (introduced), or if they grow here and are unique in some way, as by being dangerously toxic, for example, like jimsonweed (*Datura stramonium*).

This guide is, above all, a rangeland guide; therefore weedy plants found mostly on cropland and in towns are not included unless they have significantly escaped into rangeland (such as western horsenettle, *Solanum dimidiatum*). There is also an abundance of native landscape plants that grow well here (red yucca, yaupon, mountain laurel, etc.) but are not included in this book because they are not native to the rangelands of the Concho Valley.

## Geographic Area Covered

The Concho River watershed (4,308,240 acres) is the largest watershed in the Colorado River Basin (fig. 1). It is made up of the following four main watersheds:

- North Concho River watershed (965,324 acres)—drains into the North Concho River
- Middle Concho River watershed (1,696,821 acres)—drains into the Middle Concho River
- South Concho Watershed (854,649 acres)—consists of Spring Creek, Dove Creek, South Concho River, and Pecan Creek
- Concho River Watershed (791,446 acres)—drains into the Concho River

Sometimes the term "Twin Buttes watershed" is used and is meant to describe the waters leading to Twin Buttes Reservoir, which includes, collectively, the sources of the Middle Concho River, Spring Creek, Dove Creek, South Concho River, and Pecan Creek. Pecan Creek watershed, consisting solely of Pecan Creek (not shown on fig. 2), is small and for our purposes is included within the South Concho watershed. Pecan Creek is the only watershed that drains directly into Lake Nasworthy.

## Climatic Data

The Concho Valley watershed is centrally located in west central Texas and has a semiarid climate. Semiarid climates can be transitional between desert climates and humid climates, as is the case for the Concho Valley region. This means predominantly dry conditions with limited moisture. Precipitation varies from 15 to 25 inches per year (fig. 2). Although "average precipitation" is a mythical concept to local ranchers and residents, the years 1977 through 2012 demonstrated a yearly average rainfall of 21.25 inches (fig. 3). This number in itself may not sound like a bad moisture average for the local producer; however, it is not just the amount of rainfall that is significant to the local vegetation and to the producer, but the timing and distribution of this rainfall.

San Angelo annual rainfall records from 1868 to 2004 (fig. 4) reveal the extreme variation of moisture and the cyclic nature of high rainfall events. These data suggest that around every 17–21 years, there is at least one year with precipitation between 30 and 40 inches! It is important to note that although such high rainfall and runoff

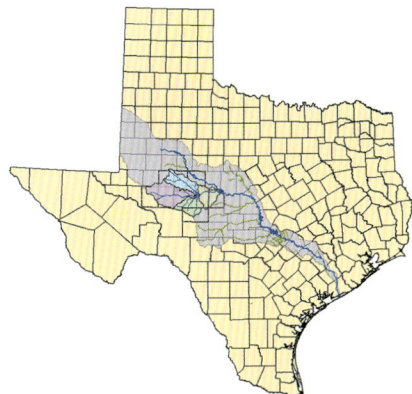

Figure 1. The Colorado River Basin. The Concho River watershed is one of nine contributing watersheds of the Colorado River Basin. The Colorado River, when it reaches the Texas coastline, eventually flows into West Matagorda Bay. ArcGIS map by George Clendenin, USDA-NRCS.

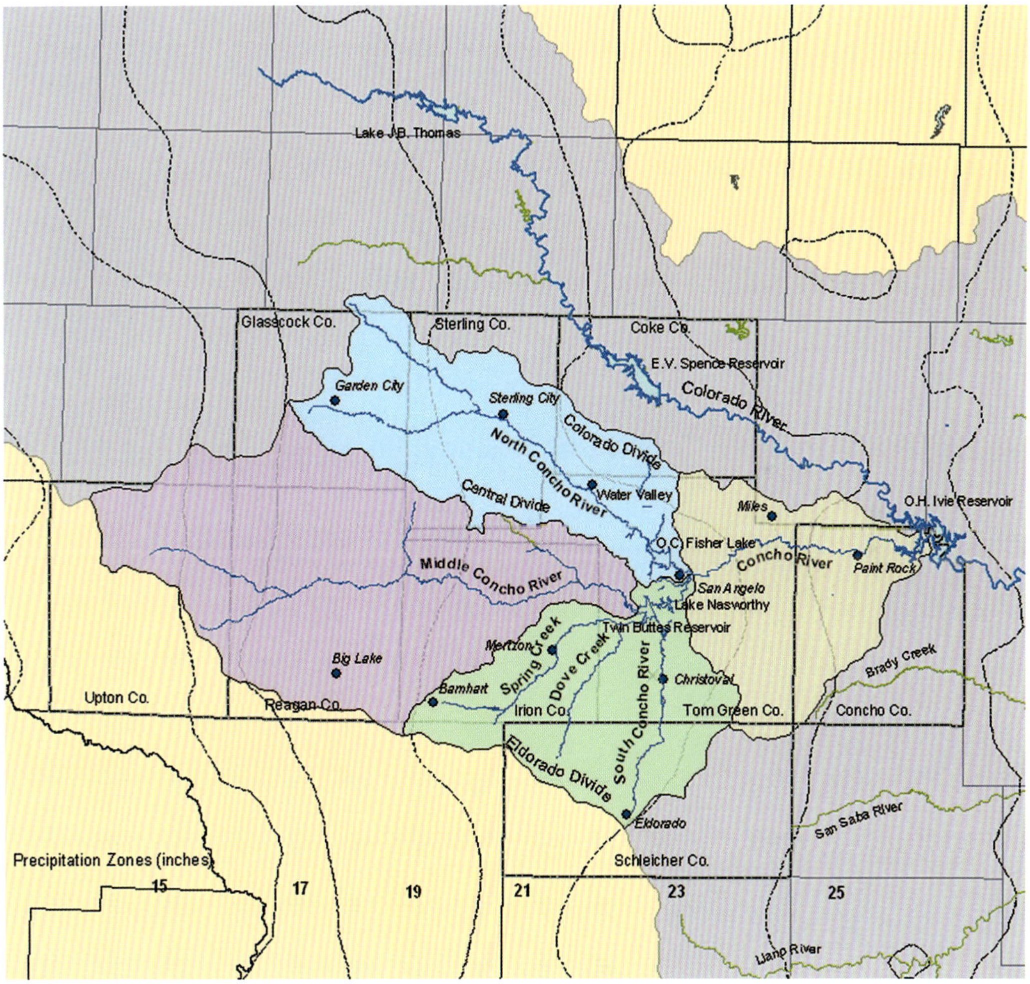

Figure 2. The Concho River watershed. This watershed receives contributed flow from over 9 West Texas counties. The map also shows precipitation zones, which vary widely between 15 and 25 inches per year. This helps us understand why the landscape transitions from short grass communities in the western part of the watershed to taller grass communities in the eastern part of the watershed. ArcGIS map by George Clendenin, USDA-NRCS.

## Historical Monthly and Annual Rainfall Data for San Angelo, Texas
Source: National Weather Service, San Angelo, Texas.

|      | JAN  | FEB  | MAR  | APR  | MAY   | JUN  | JUL  | AUG  | SEP   | OCT  | NOV  | DEC  | ANNUAL |
|------|------|------|------|------|-------|------|------|------|-------|------|------|------|--------|
| 1977 | 0.61 | 0.26 | 0.99 | 5.10 | 2.14  | 0.47 | 0.52 | 0.38 | 0.27  | 1.38 | 0.68 | 0.15 | 12.95  |
| 1978 | 0.61 | 1.17 | 0.41 | 0.73 | 1.83  | 2.81 | 0.41 | 2.93 | 1.35  | 0.42 | 1.75 | 0.25 | 14.67  |
| 1979 | 0.19 | 1.83 | 2.25 | 1.90 | 0.67  | 2.15 | 1.30 | 2.18 | 0.06  | 0.93 | 0.00 | 2.70 | 16.16  |
| 1980 | 0.84 | 0.79 | 0.71 | 0.52 | 4.72  | 3.14 | 0.33 | 3.30 | 11.00 | 0.01 | 2.53 | 2.20 | 30.09  |
| 1981 | 1.17 | 0.68 | 2.81 | 3.51 | 4.70  | 1.97 | 2.68 | 1.23 | 2.72  | 8.68 | 0.00 | 0.02 | 30.17  |
| 1982 | 1.06 | 1.53 | 0.40 | 0.83 | 4.17  | 6.01 | 0.35 | 0.46 | 0.09  | 1.26 | 1.11 | 0.91 | 18.18  |
| 1983 | 2.06 | 0.42 | 1.20 | 0.81 | 0.52  | 3.72 | 1.18 | 0.03 | 0.00  | 3.47 | 1.78 | 0.07 | 15.26  |
| 1984 | 2.38 | 0.54 | 0.49 | 0.23 | 0.54  | 2.82 | 0.60 | 0.26 | 2.99  | 3.74 | 1.09 | 3.48 | 19.16  |
| 1985 | 0.67 | 0.38 | 1.69 | 0.42 | 4.78  | 3.55 | 1.13 | 0.24 | 2.84  | 5.64 | 0.48 | 0.01 | 21.83  |
| 1986 | 0.30 | 0.65 | 0.52 | 0.07 | 7.28  | 3.30 | 0.74 | 2.50 | 7.53  | 5.72 | 1.83 | 2.48 | 32.92  |
| 1987 | 0.65 | 4.45 | 1.77 | 1.33 | 11.24 | 3.28 | 0.22 | 1.91 | 3.86  | 0.34 | 0.80 | 2.05 | 31.90  |
| 1988 | 0.01 | 0.42 | 0.69 | 1.36 | 3.31  | 2.14 | 1.22 | 0.92 | 3.20  | 0.00 | 0.00 | 0.79 | 14.06  |
| 1989 | 0.68 | 3.01 | 1.95 | 1.04 | 1.06  | 2.83 | 0.35 | 2.78 | 2.82  | 0.41 | 0.48 | 0.23 | 17.64  |
| 1990 | 1.60 | 1.65 | 0.85 | 4.14 | 4.02  | 0.05 | 4.09 | 1.05 | 6.23  | 2.40 | 2.51 | 0.21 | 28.80  |
| 1991 | 2.08 | 0.26 | 0.61 | 0.48 | 1.15  | 4.22 | 1.80 | 1.36 | 4.77  | 3.34 | 0.24 | 3.98 | 24.29  |
| 1992 | 1.90 | 3.79 | 1.05 | 1.00 | 1.49  | 4.74 | 1.98 | 1.78 | 0.27  | 1.33 | 0.95 | 0.75 | 21.03  |
| 1993 | 0.94 | 0.81 | 0.28 | 1.46 | 3.87  | 0.94 | 0.82 | 2.64 | 1.99  | 1.07 | 0.05 | 0.76 | 15.63  |
| 1994 | 1.69 | 0.34 | 0.04 | 0.97 | 3.29  | 1.04 | 0.25 | 1.30 | 3.04  | 4.27 | 1.96 | 1.21 | 19.40  |
| 1995 | 0.31 | 2.81 | 1.01 | 2.67 | 4.39  | 1.37 | 0.17 | 2.41 | 2.27  | 1.86 | 1.68 | 0.20 | 21.15  |
| 1996 | 0.06 | 0.23 | 0.28 | 2.38 | 2.08  | 1.81 | 0.26 | 7.66 | 1.92  | 2.42 | 3.35 | 0.05 | 22.50  |
| 1997 | 0.37 | 4.54 | 2.69 | 2.50 | 2.69  | 2.57 | 0.74 | 2.76 | 1.56  | 0.82 | 0.76 | 1.38 | 23.38  |
| 1998 | 0.70 | 0.53 | 1.85 | 0.00 | 1.75  | 0.88 | 0.46 | 2.77 | 0.10  | 2.39 | 1.06 | 0.49 | 12.98  |
| 1999 | 0.61 | 0.01 | 2.32 | 1.85 | 1.55  | 4.70 | 0.66 | 0.03 | 0.76  | 0.94 | 0.00 | 0.09 | 13.52  |
| 2000 | 0.08 | 0.23 | 0.77 | 0.57 | 2.21  | 3.44 | 0.02 | 0.00 | 0.58  | 3.61 | 3.08 | 0.60 | 15.19  |
| 2001 | 1.29 | 2.17 | 1.26 | 0.82 | 2.51  | 0.26 | 0.57 | 3.67 | 0.89  | 1.48 | 3.46 | 0.14 | 18.52  |
| 2002 | 0.32 | 1.10 | 1.31 | 0.33 | 0.46  | 0.88 | 2.02 | 0.41 | 1.64  | 4.06 | 0.52 | 1.37 | 14.42  |
| 2003 | 0.33 | 1.57 | 1.25 | 0.06 | 1.07  | 4.78 | 0.90 | 2.52 | 3.16  | 3.38 | 0.74 | 0.00 | 19.76  |
| 2004 | 1.37 | 1.72 | 1.70 | 1.91 | 0.86  | 3.66 | 2.18 | 4.32 | 2.05  | 5.16 | 5.18 | 0.38 | 30.49  |
| 2005 | 0.54 | 2.03 | 2.81 | 0.03 | 4.43  | 0.98 | 1.15 | 4.65 | 0.02  | 3.72 | 0.00 | 0.02 | 20.38  |
| 2006 | 0.16 | 0.64 | 1.92 | 1.34 | 1.90  | 0.30 | 0.87 | 4.87 | 2.60  | 2.21 | 0.01 | 0.83 | 17.65  |
| 2007 | 1.86 | 0.54 | 3.86 | 2.66 | 4.74  | 5.54 | 1.84 | 6.55 | 2.55  | 0.83 | 0.89 | 0.18 | 32.04  |
| 2008 | 0.39 | 0.30 | 4.64 | 0.62 | 1.01  | 2.19 | 0.19 | 3.59 | 3.99  | 1.81 | 0.23 | 0.04 | 19.00  |
| 2009 | 0.06 | 0.48 | 1.73 | 4.61 | 0.12  | 1.74 | 4.64 | 1.89 | 5.66  | 2.92 | 0.01 | 1.67 | 25.53  |
| 2010 | 2.16 | 2.72 | 1.17 | 2.65 | 1.42  | 1.96 | 1.73 | 1.10 | 1.72  | 2.51 | 0.00 | 0.99 | 20.13  |
| 2011 | 0.67 | 0.31 | 0.10 | 0.03 | 1.36  | 0.46 | 0.00 | 1.64 | 0.43  | 2.91 | 0.32 | 0.98 | 9.21   |
| 2012 | 3.30 | 2.70 | 1.49 | 0.57 | 4.50  | 0.53 | 0.52 | 0.77 | 6.91  | 0.49 | 0.00 | 0.18 | 21.96  |
|      | JAN  | FEB  | MAR  | APR  | MAY   | JUN  | JUL  | AUG  | SEP   | OCT  | NOV  | DEC  | ANNUAL |
| Normal | 0.93 | 1.35 | 1.50 | 1.42 | 2.82 | 2.59 | 1.20 | 2.26 | 2.46 | 2.73 | 1.14 | 0.85 | 21.25 |

(Normals based on the period 1981–2010, and are provided by the National Climatic Data Center)

Figure 3. Historical monthly and annual rainfall data, San Angelo. Source: National Weather Service, San Angelo, TX. (Normals are based on the period 1981–2010 and are provided by the National Climatic Data Center.)

events are very difficult for the ranching community, they are what research scientists credit for the recharge of reservoirs and larger water bodies.

I was living in Christoval, Texas, in 2004, when the area received a surprising 30 inches of rain. This was repeated in 2007, when the area received another amazing 32 inches. Our family was informed

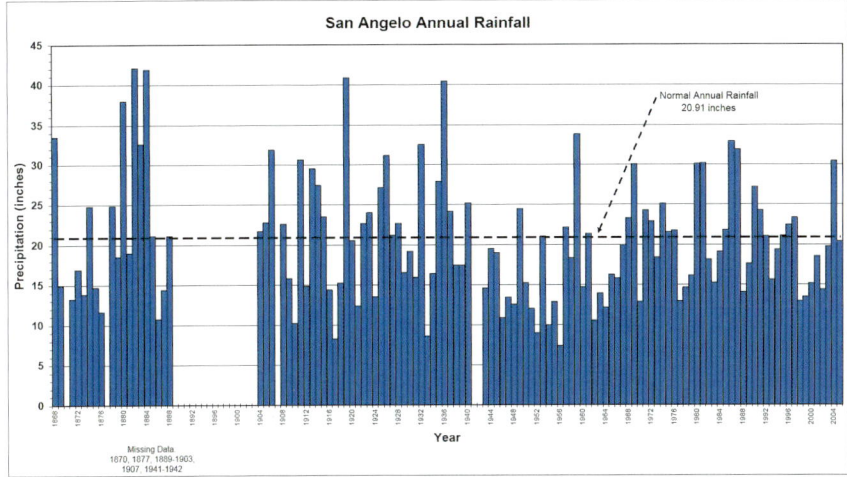

Figure 4. San Angelo annual rainfall record, 1868–2004.

Riparian area in the Concho Valley. Photo by Mark Meyer, USDA-NRCS.

several times by the local ranching community, "Don't get used to this, this is not normally how much rain we get." These were very good years for local ranchers, farmers, and producers. I also observed plant species that I had not seen in previous, drier years. An example of this was blue curls (*Phacelia congesta*). This blue-flowered plant is normally found in the Texas Hill Country and is not common in west central Texas. However, after the rains of 2007, blue curls were abundant in many sites throughout the area. This is a testament to the living seed bank that is patiently waiting for the right physical environment and moisture to burst forth and announce its presence.

Indiangrass. Photo by Dee Ann Littlefield, USDA-NRCS.

## Major Land Resource Areas of the Concho Valley

The Concho Valley can be described by the presence of two large Major Land Resource Areas, also known as MLRAs. These are the Western Edwards Plateau (MLRA 81A) and Central Edwards Plateau (MLRA 81B), with the eastern part of the valley located in the Rolling Limestone Prairie (MLRA 78A). Along the fringes of the Concho Valley are the Southern Desertic Basins, Plains, and Mountains (MLRA 42) to the west, the Southern High Plains (MLRA 77C, 77D) to the northwest, and the Rolling Red Plains (MLRA 78B, 78C) to the north (fig. 5). As you can see from this figure, most of the Concho Valley is actually located in the Edwards Plateau. While the western part of the Edwards Plateau is relatively flat, the eastern part has been eroded over time to form the hills and valleys of the current landscape.

Of particular interest for the rangeland community is why live oak grows in certain regions of the watershed and not in others. Similarly,

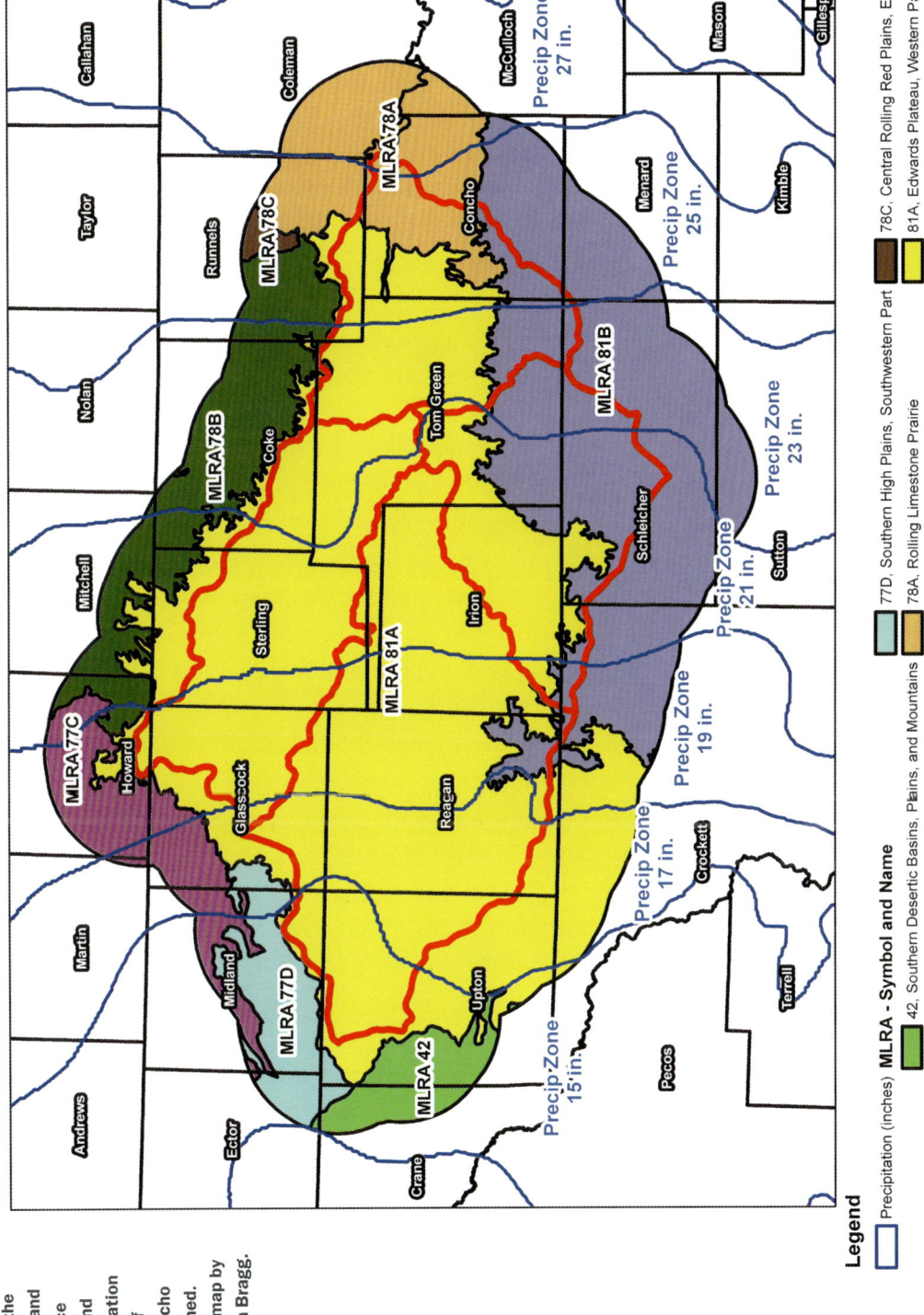

Figure 5. Map of the Major Land Resource Areas and precipitation zones of the Concho Watershed. ArcGIS map by Amanda Bragg.

why do mesquite and juniper prefer certain regions and not others? The answer lies in the relationship between the MLRA, geology, soils, and precipitation for a given area. See figure 6 for a map of the major soils.

Ector and Noelke soils are less than 20 inches to fractured limestone bedrock. In these soils, the fractures have been plugged with calcium carbonate (lime, caliche), sealing the bedrock and making it impervious to water and roots. Ector and Noelke soils have greater than 35 percent rock fragments and 20 to 35 percent clay. Ector soils occur in areas of 12 to 24 inches of average annual rainfall. Noelke soils occur in areas of 13 to 17 inches of annual rainfall and have a layer of cemented calcium carbonate directly above the limestone bedrock. Ector and Noelke soils are associated with our Ecological Site Description (ESD) called Limestone Hill. Since the fractured limestone has been sealed, live oak trees cannot thrive or grow in this environment. However, these sites are still ideal for juniper, mesquite, and various grass communities.

Talpa and Tarrant soils are less than 20 inches to fractured limestone bedrock. Talpa soils have less than 35 percent rock fragments above the limestone and 20 to 35 percent clay. Talpa soils occur in areas of 20 to 28 inches of annual rainfall. Tarrant soils have greater than 35 percent rock fragments and 40 to 60 percent clay. Tarrant soils form in areas of 20 to 34 inches of annual rainfall. Because of the increased rainfall, the fractures in the bedrock of Talpa and Tarrant soils have not been sealed with carbonates and are filled with soil, roots, and moisture. These soils are associated with our Ecological Site Description (ESD) called Low Stony Hill. Live oak trees grow on these soils because of the soil-filled fractures in the limestone. Plant roots can seek out these fractures where soil and moisture have accumulated, enabling the plants to better withstand periods of drought. These soils are primarily rangeland and are found on limestone plateaus and dissected limestone plateaus. Refer to figure 7 and the previous figures to see this relationship between MLRA, soil, plants, and precipitation. While it is not the purpose of this book to fully describe the soils of this region, I wanted to stress the importance of knowing the soils in your area and how they relate to the stewardship of the land. Contact the USDA Natural Resources Conservation Service (NRCS) to obtain further information and mapping of your specific soils.

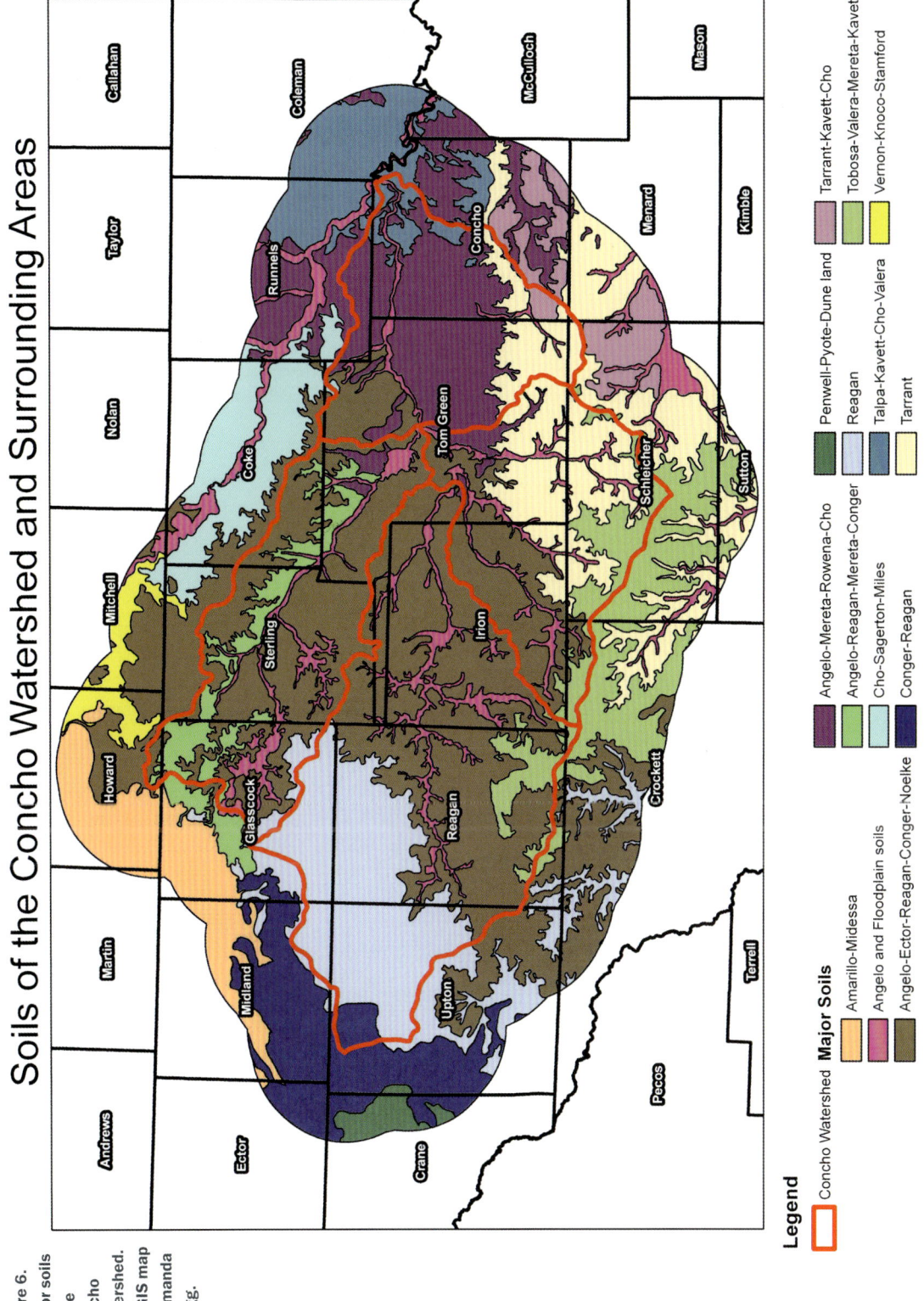

Figure 6. Major soils of the Concho Watershed. ArcGIS map by Amanda Bragg.

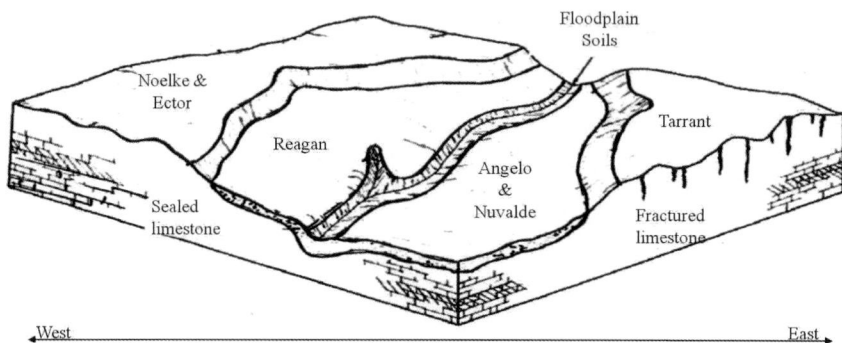

Figure 7. Block diagram of the basic soils and geology found in the Concho Watershed. As you travel from west to east in the Concho Valley, rainfall amounts increase. The soils and vegetation reflect this climate influence; short grass communities are defined in the western part, while taller grass communities grow in the eastern part. By this same measure, mesquite and juniper communities are defined in the western part, while live oak communities abound in the eastern part. Figure by USDA.

## Photographs

This book contains over 650 high-resolution pictures taken directly on ranches and roadsides of the Concho Valley. Great care was taken in selecting the final photographs; over 10,000 were taken to get to these final 650 plus! I was the photographer for most of these pictures, unless a different photographer is noted. It was a labor of love, and without fail, the most enjoyable task in this book.

One of the most important criteria for the photographs in this book was that most of them be taken in this region of west central Texas, and mostly in the Concho Valley. This was important in order to demonstrate what the plants look like in west central Texas, not in a nursery or other setting. Some plants have regional differences even though they are botanically the same species. In other words, they adapt to their environment in part because of differences in precipitation, temperature, and soil quality.

I made every effort to follow a simple formula. I strived to offer a community picture, to show what a plant looks like when walking up to it; a shot of the individual plant, which offers a little more detail; and a close-up, which reveals identifiable features such as a flower or leaf shape. I feel this is a practical approach to identifying plants in the field.

## Plant Names

Plants are classified based on morphological (shape), cytological (cellular), anatomical (structural makeup), biochemical (chemical reactions in living organisms), and ecological (environmental) characteristics. In other words, plants are grouped based on like characteristics that set them apart from other groups. The realm of categorizing and naming plants is handled by taxonomists. Taxonomists and plant biologists use scientific plant names that are composed of three parts: the genus, specific epithet, and authority(ies). Taxonomists do not use common names because they can be general or regional or can refer to several plants; they can thereby be misleading. However, scientific names can change when new information comes to light. It can sometimes be frustrating to search for a plant that has had its scientific name changed.

In this book, plants are listed by a common name used in the area, combined with a two-part scientific name. If the scientific name of a plant has changed fairly recently, the synonym, or what the plant name used to be, will be shown following the current scientific name with the abbreviation "Sy," meaning synonym. I have used *Vascular Plants of Texas* (Jones, Wipff, and Montgomery 1997) for this valuable bread-crumb trail. I have also used this source for the nomenclature (names of plants and their families) of nongrass plants, as well as *Shinners and Mahler's Illustrated Flora of North Central Texas* (Diggs, Lipscomb, and O'Kennon 1999). In general, these two sources agree on most of the scientific names on our plant list. When they disagree, a name is chosen based on the latest published source of information. Since the Diggs, Lipscomb, and O'Kennon volume is primarily a guide to north-central Texas, several taxa in this west central Texas book are not included in that original work. In those cases, the primary reference was Jones, Wipff, and Montgomery (1997). Nomenclature for grasses follows predominantly the names used in the Grasses of Texas Checklist, August 3, 2009, by Dr. Stephan Hatch, Texas A&M University, with some exceptions used by the USDA.

### Toxic/Poisonous Plants

Toxic plant information was derived primarily from *Toxic Plants of Texas* (Hart et al. 2010). It is important to note that toxic plants are frequently found on rangelands and may serve various ecological purposes. Toxic plant poisonings of grazing animals usually occur in situations of overgrazing and/or drought, where the toxic plant may be one of few choices left in the pasture.

Research on poisonings has determined that foraging behavior is often learned from the parent. In other words, the offspring eat what the mother eats, and the mother knows to stay away from certain plants. Therefore, poisonings may occur when new animals are brought into a pasture or ranch with which they are not familiar.

In addition, a complex relationship exists between the animal and the plant in order for toxic poisonings to occur. For the animal, factors such as its species, class, age, nutritional health, live weight, and genetics need to be considered, while for the plant, factors such as its type, stage of growth, which part is eaten, what quantity is eaten, and the time period of repeated exposure all play into the equation.

To add to the complexity, some plants seem toxic only in certain geological areas and soils and not in others. Broom snakeweed (*Gutierrezia sarothrae*) may accumulate selenium when growing on high-selenium soil, but it is relatively nontoxic when growing on clay soils. Another example is live oak (*Quercus virginiana*), shin oak (*Quercus sinuata* var. *breviloba*), or any of the *Quercus* species listed here. Although oak is marked with a *T* and listed in Hart et al. (2010) as a toxic plant, residents of the Concho Valley sometimes feed their goats a moderate number of oak branches, which they readily consume with no apparent problems.

This all leads to one human consideration: don't panic if your land has a plant with a *T* beside its name in the plant list, unless of course it is the only plant in your pasture.

### Noxious and Invasive Plants

In plant descriptions, the terms "noxious," "invasive," and "introduced" have often mistakenly been used interchangeably. However, these terms have specific definitions for legal purposes. Executive Order 13112 was issued on February 3, 1999, to enhance federal coordination and response to the complex and accelerating problem of invasive species. An "invasive" plant, by definition, is one that is likely to cause economic or environmental harm or harm to human health. A plant is "noxious" when it is so designated by federal or state declaration. A plant may be a "pest," but that does not mean it is noxious by state or federal definition. In addition, just because a plant is introduced does not mean that it is harmful or invasive or noxious. There are many introduced plants that are beneficial in the United States.

There is a Federal List of Noxious Plants as well as individual state lists. State lists include noxious plants that are identified and described by the individual state legislatures and their laws and policies. There are also federal policies set by departments and agencies

Figure 8.
Texas State-Listed Noxious Plants, Texas Administrative Code (TAC), Title 4, Part I, Chapter 19, Subchapter T, Rule §19.300, Noxious and Invasive Plant List (adopted 2005, amended June 10, 2007).

| Common Name | Botanical Name |
|---|---|
| **Noxious plants** | |
| alligatorweed | *Alternanthera philoxeroides* |
| balloonvine | *Cardiospermum halicacabum* |
| Brazilian peppertree | *Schinus terebinthifolius* |
| broomrape | *Orobanche ramosa* |
| camelthorn | *Alhagi camelorum* |
| Chinese tallow tree | *Triadica sebifera* |
| Eurasian watermilfoil | *Myriophyllum spicatum* |
| giant duckweed | *Spirodela oligorrhiza* |
| giant reed | *Arundo donax* |
| hedge bindweed | *Calystegia sepium* |
| hydrilla | *Hydrilla verticillata* |
| itchgrass | *Rottboellia cochinchinensis* |
| Japanese dodder | *Cuscuta japonica* |
| kudzu | *Pueraria montana* var. *lobata* |
| lagarosiphon | *Lagarosiphon major* |
| paperbark | *Melaleuca quinquenervia* |
| purple loosestrife | *Lythrum salicaria* |
| rooted waterhyacinth | *Eichhornia azurea* |
| saltcedar | *Tamarix* spp. |
| salvinia | *Salvinia* spp. |
| serrated tussock | *Nassella trichotoma* |
| torpedograss | *Panicum repens* |
| tropical soda apple | *Solanum viarum* |
| water spinach | *Ipomoea aquatica* |
| waterhyacinth | *Eichhornia crassipes* |
| waterlettuce | *Pistia stratiotes* |
| **Invasive plants** | |
| Chinese tallow tree | *Triadica sebifera* |
| kudzu | *Pueraria montana* var. *lobata* |
| saltcedar | *Tamarix* spp. |
| tropical soda apple | *Solanum viarum* |

such as the USDA and the NRCS to further define the problems and set forth plans of action. All these lists share one thing in common: the plants were usually introduced into an area and have the potential to do serious harm to the economy and the environment.

The state of Texas adopted legislation in 2005 (amended in 2007) to list and address noxious plants for Texas. This list contains 26 plant species identified to be noxious in the state of Texas (fig. 8).

Figure 9.
State-listed noxious plants of our neighboring states.

| Neighboring State | Number of State-Listed Noxious Plants |
|---|---|
| Arkansas | 38 |
| Colorado | 72 |
| Kansas | 12 |
| Louisana | 1 |
| New Mexico | 32 |
| Oklahoma | 3 |

Yellow star-thistle (*Centaurea solstitialis*, left) and malta star-thistle (*C. melitensis*, right) are starting to appear in the Concho Valley and are increasing. These photos were taken along roadsides in the San Angelo city limits. Star-thistle species do best in summer drought conditions, often overtaking native plants in abundance. Although it is not state-listed yet, it is an invasive plant not to be ignored. Photos by George Clendenin, USDA-NRCS.

### Noxious Plants in Neighboring States

Why is it important to know the "weeds" of our neighbors? We have plants appearing in the Concho Valley that are already on noxious plant lists in neighboring states, and we should be alert to their presence. An example is star-thistles (*Centaurea* spp.). Both Malta and yellow star-thistles are beginning to appear in the Concho Valley and are already listed as noxious in 12 other states (fig. 9).

### Noxious Plant Definitions

Here are some definitions taken directly from the NRCS Invasive Species Policy (and similar to the definitions given in Executive Order 13112), which will be helpful in understanding the concepts involving noxious and invasive species.

Ecosystem—The complex of a community of organisms and its environment.

Introduction—The intentional or unintentional escape, release, dissemination, or placement of a species into an ecosystem as a result of human activity. "Introduced" is not synonymous with the term "invasive" and should not be confused with it.

Invasive plant—A species whose introduction causes, or is likely to cause, economic or environmental harm or harm to human health.

Native species—Within a particular ecosystem, a species that, other than as a result of an introduction, historically occurred or currently occurs in that ecosystem.

Nonnative species—Within a particular ecosystem, any species— including its seeds, eggs, spores, or other biological material capable of propagating that species—that is not native to that ecosystem.

Noxious plant—Any plant species designated as such by the secretary of agriculture, secretary of the interior, or by state law or regulation. Generally, noxious weeds will be one or more of the following: (1) aggressive and difficult to manage, (2) parasitic, (3) a carrier or host of deleterious insects or disease, or (4) nonnative in, new to, or not common in the United States or parts thereof.

Pest—A weed, insect, disease, animal, or other organism (including invasive and noninvasive species) that directly or indirectly causes damage or annoyance by destroying food and fiber products, causing structural damage, or creating a poor environment for other organisms.

## Ethnobotany of Plants

Ethnobotany is the scientific study of the relationships between people (ethno) and plants (botany). Quite simply, it is the study of how various peoples used plants historically and how they still potentially use them today. For our purposes, ethnobotany as covered in this book deals primarily with plant use by early Native Americans and pioneering settlers. This book is not intended to be an exhaustive study on the ethnobotanical uses of its featured plants. However, ethnobotanical information is sometimes provided for anecdotal purposes and better appreciation for the usefulness of some plants. I do not intend or recommend that the reader try any of the historic uses of these plants, including their application for medicinal purposes.

## References and Literature Cited

Although it was not my original intention, I made every effort to include references and bibliographical information for the sources cited in this book. The references are meant to be inclusive and not exclusive, and I hope they will offer a springboard for your continued journey in land stewardship. I highly encourage you to seek out this information and these resources. These references have been an invaluable addition to my own personal library and growth and may be for you as well.

The author, George Clendenin, at work in the field. Photo by Dee Ann Littlefield, USDA-NRCS.

Maximilian sunflower. Photo by Dee Ann Littlefield, USDA-NRCS.

# Section One

# Common Rangeland Plant List

The following common rangeland plants are found in the Concho Valley and west central Texas and are described in this book.

Key to plant symbols:

Origin: ▲—Introduced into the United States (otherwise assume native)

*T*—Plant has toxic properties or may affect the health of grazing animals, either directly or indirectly, as documented by *Toxic Plants of Texas* (Hart et al. 2010).

## Grasses and Grasslike Plants

|   | Common name | Scientific name |
|---|---|---|
| | CYPERACEAE—**Sedge Family** | |
| | Cedar sedge | *Carex planostachys* |
| | Spike rush | *Eleocharis* spp. |
| | LILIACEAE—**Lily Family** | |
| *T* | Wild onion | *Allium drummondii* |
| *T* | Rain lily | *Cooperia drummondii* |
| | POACEAE—**Grass Family** | |
| | Big bluestem | *Andropogon gerardii* |
| | Bushy bluestem | *Andropogon glomeratus* |
| | Red threeawn | *Aristida purpurea* var. *longiseta* |
| | Wright threeawn | *Aristida purpurea* var. *wrightii* |
| | Silver bluestem | *Bothriochloa laguroides* |
| | Sideoats grama | *Bouteloua curtipendula* |
| | Hairy grama | *Bouteloua hirsuta* |
| | Texas grama | *Bouteloua rigidiseta* |
| | Red grama | *Bouteloua trifida* |
| ▲ | Rescue grass | *Bromus catharticus* |
| | Buffalograss | *Buchloe dactyloides* |
| | Hooded windmillgrass | *Chloris cucullata* |
| | Tumble windmillgrass | *Chloris verticillata* |
| | Fall witchgrass | *Digitaria cognata* |
| | Canada wildrye | *Elymus canadensis* |
| | Squirreltail | *Elymus longifolius* |
| | Plains lovegrass | *Eragrostis intermedia* |
| | Texas cupgrass | *Eriochloa sericea* |
| | Hairy tridens | *Erioneuron pilosum* |
| | Curly-mesquite | *Hilaria belangeri* |
| | Green sprangletop | *Leptochloa dubia* |
| | Three-flower melic | *Melica nitens* |
| | Texas wintergrass | *Nassella leucotricha* |
| | Hall panicum | *Panicum hallii* |
| | Vine-mesquite | *Panicum obtusum* |
| | Switchgrass | *Panicum virgatum* |
| | Western wheatgrass | *Pascopyrum smithii* |
| | Knotgrass | *Paspalum distichum* |
| | Tobosagrass | *Pleuraphis mutica* |

| | Little bluestem | *Schizachyrium scoparium* |
| --- | --- | --- |
| | Plains bristlegrass | *Setaria leucopila* |
| | Reverchon bristlegrass | *Setaria reverchonii* |
| | Yellow Indiangrass | *Sorghastrum nutans* |
| | Tall dropseed | *Sporobolus compositus* |
| | Sand dropseed | *Sporobolus cryptandrus* |
| | White tridens | *Tridens albescens* |
| | Slim tridens | *Tridens muticus* var. *muticus* |
| | Eastern gamagrass | *Tripsacum dactyloides* |

**Forbs**

| | Common name | Scientific name |
| --- | --- | --- |
| ACANTHACEAE — **Wild Petunia Family** | | |
| | Ruellia, wild petunia | *Ruellia metziae* |
| | Hairy tube tongue | *Siphonoglossa pilosella* |
| APIACEAE — **Carrot or Parsley Family** | | |
| | Purple eryngo | *Eryngium leavenworthii* |
| ARISTOLOCHIACEAE — **Pipevine or Birthwort Family** | | |
| | Dutchman's pipe | *Aristolochia coryi* |
| ASCLEPIADACEAE — **Milkweed Family** | | |
| T | Antelope horn milkweed | *Asclepias asperula* |
| T | Broadleaf milkweed | *Asclepias latifolia* |
| ASTERACEAE — **Sunflower or Daisy Family** | | |
| | Huisache daisy | *Amblyolepis setigera* |
| | Field ragweed | *Ambrosia confertiflora* |
| | Western ragweed | *Ambrosia psilostachya* |
| | Lazy daisy | *Aphanostephus skirrhobasis* |
| | Mexican sagewort | *Artemisia ludoviciana* |
| | Heath aster | *Aster ericoides* |
| | Engelmann daisy | *Engelmannia peristenia* |
| | Fleabane | *Erigeron strigosus* |
| | Rabbit tobacco | *Evax verna* |
| | Pincushion daisy | *Gaillardia pinnatifida* |
| | Indian blanket | *Gaillardia pulchella* |
| | Curlycup gumweed | *Grindelia nuda* |
| | Sawleaf daisy | *Grindelia papposa* |
| | Common broomweed broomweed | *Gutierrezia dracunculoides* |
| T | Broom snakeweed | *Gutierrezia sarothrae* |
| | Sticky selloa | *Gymnosperma glutinosum* |
| | Gray golden aster | *Heterotheca canescens* |
| | Chalk Hill woolly-white | *Hymenopappus artemisiifolius* |
| | Gayfeather | *Liatris punctata* |
| | Texas skeleton plant | *Lygodesmia texana* |
| | Cutleaf daisy | *Machaeranthera pinnatifida* |
| | Rock daisy | *Melampodium leucanthum* |
| | Lyreleaf parthenium | *Parthenium confertum* |
| | Upright prairie coneflower | *Ratibida columnifera* |
| T | Threadleaf groundsel | *Senecio douglasii* |
| | Bush sunflower | *Simsia calva* |

|   |   |   |
|---|---|---|
|   | Tall goldenrod | *Solidago canadensis* |
|   | Greenthread | *Thelesperma filifolium* |
|   | Common dogweed | *Thymophylla pentachaeta* |
|   | Cowpen daisy | *Verbesina encelioides* |
|   | Orange zexmenia | *Wedelia texana* |
|   | Plains zinnia | *Zinnia grandiflora* |

**BORAGINACEAE — Borage Family**

|   |   |   |
|---|---|---|
|   | Rat-ear coldenia | *Tiquilia canescens* |

**BRASSICACEAE — Mustard Family**

|   |   |   |
|---|---|---|
|   | Pepperweed | *Lepidium virginicum* |

**CAMPANULACEAE — Bluebell or Bellflower Family**

|   |   |   |
|---|---|---|
| T | Cardinal flower | *Lobelia cardinalis* |

**CHENOPODIACEAE — Goosefoot Family**

|   |   |   |
|---|---|---|
| T ▲ | Tumbleweed | *Salsola tragus* |

**EUPHORBIACEAE — Euphorbia or Spurge Family**

|   |   |   |
|---|---|---|
|   | Lindheimer's copperleaf | *Acalypha phleoides* |
|   | Hoary euphorbia | *Chamaesyce lata* |
|   | Grassland croton | *Croton dioicus* |
|   | One-seed croton | *Croton monanthogynus* |
|   | Leatherweed croton | *Croton pottsii* |
|   | Low wild mercury | *Ditaxis humilis* |
| T | Snow-on-the-mountain | *Euphorbia marginata* |
|   | Knotweed leaf-flower | *Phyllanthus polygonoides* |
| T | Texas stillingia | *Stillingia texana* |
| T | Trecul stillingia | *Stillingia treculiana* |
|   | Noseburn | *Tragia ramosa* |

**FABACEAE — Legume or Bean Family**

|   |   |   |
|---|---|---|
|   | Nuttall peavine | *Astragalus nuttallianus* |
|   | Purple dalea | *Dalea lasiathera* |
|   | Dwarf dalea | *Dalea nana* |
|   | Velvet bundleflower | *Desmanthus velutinus* |
|   | Texas bluebonnet | *Lupinus subcarnosus* |
| ▲ | Bur clover | *Medicago minima* |
|   | Sensitive briar | *Mimosa nuttallii* |
|   | Texas snoutbean | *Rhynchosia senna* |
| T | Twin-leaf senna | *Senna roemeriana* |
|   | Deer pea vetch | *Vicia ludoviciana* |

**GENTIANACEAE — Gentian Family**

|   |   |   |
|---|---|---|
| T | Mountain pink | *Centaurium beyrichii* |

**GERANIACEAE — Geranium Family**

|   |   |   |
|---|---|---|
| ▲ | California filaree | *Erodium cicutarium* |
|   | Texas filaree | *Erodium texanum* |

**KRAMERIACEAE — Ratany Family**

|   |   |   |
|---|---|---|
|   | Trailing ratany | *Krameria lanceolata* |

**LAMIACEAE — Mint Family**

|   |   |   |
|---|---|---|
|   | False pennyroyal | *Hedeoma drummondii* |
| ▲ | Common horehound | *Marrubium vulgare* |
|   | Lemon beebalm | *Monarda citriodora* |
|   | Mealycup sage | *Salvia farinacea* |

| | | |
|---|---|---|
| *T* | Lance-leaf sage | *Salvia reflexa* |
| | Texas salvia | *Salvia texana* |
| | Skullcap | *Scutellaria drummondii* |

**LOASACEAE — Stickleaf or Blazingstar Family**

| | | |
|---|---|---|
| | Stickleaf | *Mentzelia oligosperma* |

**MALVACEAE — Mallow Family**

| | | |
|---|---|---|
| | Indian mallow | *Abutilon fruticosum* |
| | Wine cup | *Callirhoe involucrata* |
| | Bladderpod sida | *Rhynchosida physocalyx* |
| | Spreading sida | *Sida abutifolia* |
| | Narrowleaf globemallow | *Sphaeralcea angustifolia* |
| | Scarlet globemallow | *Sphaeralcea coccinea* |

**NYCTAGINACEAE — Four-o'clock Family**

| | | |
|---|---|---|
| | Angel trumpets | *Acleisanthes longiflora* |
| | Narrowleaf spiderling | *Boerhavia linearifolia* |
| | Scarlet musk flower | *Nyctaginia capitata* |

**OLEACEAE — Olive Family**

| | | |
|---|---|---|
| | Low menodora | *Menodora heterophylla* |

**ONAGRACEAE — Evening Primrose Family**

| | | |
|---|---|---|
| | Western primrose | *Calylophus hartwegii* |
| | Limestone gaura | *Gaura calcicola* |
| | Showy evening primrose | *Oenothera speciosa* |

**OXALIDACEAE — Woodsorrel Family**

| | | |
|---|---|---|
| | Drummond's oxalis | *Oxalis drummondii* |
| | Woodsorrel | *Oxalis stricta* |

**PAPAVERACEAE — Poppy Family**

| | | |
|---|---|---|
| | White pricklypoppy | *Argemone aurantiaca* |

**PHYTOLACCACEAE — Pokeweed Family**

| | | |
|---|---|---|
| | Pigeonberry | *Rivina humilis* |

**PLANTAGINACEAE — Plantain Family**

| | | |
|---|---|---|
| | Heller's plantain | *Plantago helleri* |
| | Red-seed plantain | *Plantago rhodosperma* |

**POLYGALACEAE — Milkwort Family**

| | | |
|---|---|---|
| | White milkwort | *Polygala alba* |

**RUBIACEAE — Coffee or Madder Family**

| | | |
|---|---|---|
| | Bluets | *Hedyotis nigricans* |

**RUTACEAE — Citrus Family**

| | | |
|---|---|---|
| *T* | Dutchman's breeches | *Thamnosma texana* |

**SCROPHULARIACEAE — Figwort Family**

| | | |
|---|---|---|
| | Penstemon | *Penstemon cobaea* |

**SOLANACEAE — Nightshade Family**

| | | |
|---|---|---|
| *T* | Jimsonweed | *Datura stramonium* |
| | Ground cherry | *Physalis cinerascens* |
| *T* | Western horsenettle | *Solanum dimidiatum* |
| *T* | Silverleaf nightshade | *Solanum elaeagnifolium* |
| *T* | Buffalobur | *Solanum rostratum* |
| | Texas nightshade | *Solanum triquetrum* |

**VERBENACEAE — Vervain or Verbena Family**

| | | |
|---|---|---|
| | Prairie verbena | *Glandularia bipinnatifida* |
| | Pink verbena | *Glandularia pumila* |

|  |  |
|---|---|
| Texas frogfruit | *Phyla nodiflora* |
| Slender vervain | *Verbena halei* |

**ZYGOPHYLLACEAE** — **Caltrop Family**

*T* ▲ Goathead, puncturevine — *Tribulus terrestris*

## Shrubs and Trees

| Common name | Scientific name |
|---|---|

**ANACARDIACEAE** — **Sumac Family**

| Flameleaf sumac | *Rhus copallinum* |
|---|---|
| Littleleaf sumac | *Rhus microphylla* |
| Skunkbush sumac | *Rhus trilobata* |
| Evergreen sumac | *Rhus virens* |

**ASTERACEAE** — **Sunflower or Daisy Family**

| Willow baccharis | *Baccharis neglecta* |
|---|---|

**BERBERIDACEAE** — **Holly Family**

| Algerita | *Berberis trifoliolata* |
|---|---|

**CAPRIFOLIACEAE** — **Honeysuckle Family**

| White honeysuckle | *Lonicera albiflora* |
|---|---|

**CHENOPODIACEAE** — **Goosefoot Family**

| Fourwing saltbush | *Atriplex canescens* |
|---|---|

**CUPRESSACEAE** — **Cypress Family**

| Ashe juniper (blueberry) | *Juniperus ashei* |
|---|---|
| Redberry juniper | *Juniperus pinchotii* |

**EBENACEAE** — **Persimmon or Ebony Family**

*T* Texas persimmon — *Diospyros texana*

**EPHEDRACEAE** — **Mormon Tea Family**

| Ephedra, Mormon tea | *Ephedra antisyphilitica* |
|---|---|

**FABACEAE** — **Legume or Bean Family**

|  |  |
|---|---|
| Catclaw acacia | *Acacia greggii* |
| Roemer's acacia | *Acacia roemeriana* |
| Feather dalea | *Dalea formosa* |
| Catclaw mimosa | *Mimosa aculeaticarpa* var. *biuncifera* |
| Fragrant mimosa | *Mimosa borealis* |

*T* Honey mesquite — *Prosopis glandulosa*

**FAGACEAE** — **Oak Family**

| | | |
|---|---|---|
| *T* | Bur oak | *Quercus macrocarpa* |
| *T* | Mohr shin oak | *Quercus mohriana* |
| *T* | Vasey shin oak | *Quercus pungens* var. *vaseyana* |
| *T* | White shin oak | *Quercus sinuata* var. *breviloba* |
| *T* | Live oak | *Quercus virginiana* |

**JUGLANDACEAE** — **Walnut Family**

| Pecan | *Carya illinoinensis* |
|---|---|
| Little walnut | *Juglans microcarpa* |

**OLEACEAE** — **Olive Family**

| Elbowbush | *Forestiera pubescens* |
|---|---|

**RHAMNACEAE** — **Buckthorn Family**

| Javelina bush | *Condalia ericoides* |
|---|---|
| Green condalia | *Condalia hookeri* |
| Lotebush | *Ziziphus obtusifolia* |

RUBIACEAE — **Coffee or Madder Family**
T	Buttonbush	*Cephalanthus occidentalis*
RUTACEAE — **Citrus Family**
	Tickle tongue	*Zanthoxylum hirsutum*
SALICACEAE — **Willow Family**
	Black willow	*Salix nigra*
SAPINDACEAE — **Soapberry Family**
	Western soapberry	*Sapindus saponaria* var. *drummondii*
SAPOTACEAE — **Chicle, Sapodilla, or Sapote Family**
	Spiny bumelia	*Sideroxylon lanuginosum*
SOLANACEAE — **Nightshade Family**
	Wolfberry	*Lycium berlandieri*
TAMARICACEAE — **Salt Cedar Family**
▲	Salt cedar	*Tamarix gallica*
ULMACEAE — **Elm Family**
	Netleaf hackberry	*Celtis laevigata* var. *reticulata*
	American elm	*Ulmus americana*
VERBENACEAE — **Vervain or Verbena Family**
T	Whitebrush	*Aloysia gratissima*
VISCACEAE — **Mistletoe Family**
T	Mistletoe	*Phoradendron tomentosum*

## Vines

	Common name	Scientific name
ANACARDIACEAE — **Sumac Family**
	Poison ivy	*Toxicodendron radicans*
ASCLEPIADACEAE — **Milkweed Family**
	Wavy-leaf milkvine	*Funastrum crispum*
CONVOLVULACEAE — **Morning Glory Family**
▲	Field bindweed	*Convolvulus arvensis*
	Texas bindweed	*Convolvulus equitans*
	Cotton morning glory	*Ipomoea cordatotriloba*
	Lindheimer morning glory	*Ipomoea lindheimeri*
CUCURBITACEAE — **Gourd Family**
	Buffalo gourd	*Cucurbita foetidissima*
	Balsam gourd	*Ibervillea lindheimeri*
MENISPERMACEAE — **Moonseed Family**
	Carolina snailseed	*Cocculus carolinus*
RANUNCULACEAE — **Buttercup Family**
	Old man's beard	*Clematis drummondii*
SMILACACEAE — **Greenbriar Family**
	Greenbriar	*Smilax bona-nox*
VITACEAE — **Grape Family**
	Ivy treebine	*Cissus incisa*
	Virginia creeper	*Parthenocissus quinquefolia*

## Cacti and Agaves

|   | Common name | Scientific name |
|---|---|---|
|   | AGAVACEAE—Century-Plant Family | |
| T | Sacahuista | *Nolina texana* |
|   | Buckley yucca | *Yucca constricta* |
|   | CACTACEAE—Cactus Family | |
|   | Horse crippler | *Echinocactus texensis* |
|   | Strawberry cactus | *Echinocereus enneacanthus* |
|   | Lacy cactus | *Echinocereus reichenbachii* |
|   | Prickly pear | *Opuntia engelmannii* |
|   | Tasajillo | *Opuntia leptocaulis* |

# Plant Profiles

Sedges look very much like grass, with long, slender leaves emerging from a central clump. At first glance, you may wince at this photo, thinking that you are looking at a grass bur, but you are not! The fruit or seedhead of cedar sedge is not "armed" or prickly. Photos by George Clendenin, USDA-NRCS.

GRASSES AND GRASSLIKE PLANTS

# Cedar sedge *(Carex planostachys)*
CYPERACEAE — SEDGE FAMILY

**Characteristics:** Native; perennial; cool-season.

**Growth Form:** Short, bunch-forming sedge; height 4–8 inches including seedhead.

**Reproduction:** Seeds and short rhizomes.

**General Description:** Most sedges are found in wet places, but this species grows in dry, shallow, rocky soil, often in association with live oak or juniper. It prefers partial shade. This species is essentially evergreen. Leaves are shiny and slick and a different color and texture than grass leaves. A short, small seedhead composed of a cluster of seeds is produced in early spring.

**Livestock and Wildlife Value:** Provides good forage for livestock and white-tailed deer. This plant is especially valuable in winter and early spring when grasses are dormant and green forage is needed. Seedheads are eaten by turkeys in late winter and early spring.

**Management:** Cedar sedge is maintained by good conservative livestock grazing and control of deer numbers. In a dry winter, it is often heavily grazed and will benefit greatly from a grazing rest in the spring. It often exists under the canopy of oak, juniper, or other shrubs where it is protected from heavy grazing. After mechanical brush control, this plant often increases.

Photos by
George Clendenin,
USDA-NRCS.

# Spike rush *(Eleocharis* spp.*)*
## CYPERACEAE—SEDGE FAMILY

**Characteristics:** Native; perennial; warm-season.

**Growth Form:** Colony-forming sedge; height 8–24 inches depending on species.

**Reproduction:** Seeds and extensive rhizomes.

**General Description:** There are 2 or 3 common species of spike rush in the Concho Valley, which are usually lumped together for practical management purposes. All species grow near water, often at the edge of creeks, seeps, springs, or ponds. Plants consist of many round, slender, hollow stems with a small spiked seedhead at the tip. Other common name: spike sedge.

**Livestock and Wildlife Value:** Occasionally grazed by livestock and white-tailed deer, but not preferred. Ducks and shorebirds commonly eat the seeds and sometimes graze the stems.

**Management:** Spike rush is valuable for helping to stabilize creek banks and riparian areas. The dense network of roots and interconnected rhizomes provides good binding of creekside soils and helps prevent erosion during flooding. Research indicates that spike rush can produce over 20 miles of root in 1 cubic foot of soil. Grazing management in riparian areas should be directed toward maintaining a dense cover of plants to dissipate the energy of floodwater, reduce erosion, and protect banks.

Photos by George Clendenin, USDA-NRCS.

# Wild onion *(Allium drummondii)*
## LILIACEAE—LILY FAMILY

**Characteristics:** Native; perennial from bulb.

**Growth Form:** Weakly upright plant from small bulb; height 6–12 inches.

**Reproduction:** Seeds and/or movement of bulbs by burrowing animals.

**General Description:** Wild onion and wild garlic are the same plant (*A. drummondii*) and have white and/or pink to purple flowers. Wild onion stems and leaves are fleshy, similar to those of domestic onions, only much smaller. The small onions have a strong onion or garlic scent and are covered by a layer of fibrous mesh. Other common names: wild garlic, Drummond's onion, prairie onion.

**Livestock and Wildlife Value:** Grazed in early spring by livestock and white-tailed deer. Bulbs are sometimes rooted by turkeys, javelinas, feral hogs, and small rodents. Seeds may have some value to songbirds. Cows grazing on wild onions are known to give onion-flavored milk and butter.

**Management:** No specific management is needed to maintain modest populations of this plant. Good range management practices that provide adequate forage will prevent wild onion poisonings.

**Toxic Agent:** N-propyl disulfide, which destroys red blood cells.

**Toxic Description:** Cattle and horses are susceptible to onion poisoning, and cats are very sensitive to it. Sheep are more resistant but have been poisoned in some instances. Usually long-term intake or excessive consumption is required for poisonings. See *Toxic Plants of Texas* (Hart et al. 2010) for livestock clinical signs and more information.

Flower—The large, white to pink-tinted tubular flowers have 6 pointed petals, distinguishing them from all other flowers in the Concho Valley.

Seedpod—After flowers are pollinated, a 3-lobed seedpod is formed.

Photos by George Clendenin, USDA-NRCS.

# Rain lily *(Cooperia drummondii)*
## LILIACEAE — LILY FAMILY

**Characteristics:** Native; perennial from bulb.

**Growth Form:** Thin, upright forb growing from an underground bulb approximately 1 inch in diameter; flowering height about 12 inches.

**Reproduction:** Seeds.

**General Description:** Flower stalks of rain lily grow quickly within a few days of significant rainfall. A few succulent, grasslike leaves may be present during the entire growing season but often go unnoticed until the flowers emerge. Seeds are large, flat, and black.

**Livestock and Wildlife Value:** Leaves and flower stalks are readily eaten by livestock and white-tailed deer. Seedpods are likely eaten by turkeys prior to shattering.

**Management:** Proper conservative grazing is the only management practice needed to maintain this species. Bulbs can be dug and transplanted by those wishing to add this plant to wildflower gardens.

**Toxic Agent:** Not yet identified but found only in the dead leaf tips of plants growing in a limited geographical area. Photosensitization (sunburn) usually occurs within 10 days after a significant rainfall has occurred on dead leaves.

**Toxic Description:** Rain lily causes primarily photosensitization, and poisoning results in loss of production but usually not death. Although toxicity has not been a problem in the Concho Valley, highly photodynamic activity has occurred in DeWitt, Gonzales, and Caldwell Counties. See *Toxic Plants of Texas* (Hart et al. 2010) for livestock clinical signs and more information.

**Seedhead**—The seedhead gives rise to the alternate common name "turkeyfoot," because of its shape and color. The branched seedhead can have 2–6 branches, although 3 are common. Photos by George Clendenin, USDA-NRCS.

# Big bluestem *(Andropogon gerardii)*
## POACEAE — GRASS FAMILY

**Characteristics:** Native; perennial; warm-season.

**Growth Form:** Tall, robust bunchgrass, but big bluestem in the Concho Valley is typically more rhizomatous and colony forming; mature height 3–6 feet.

**Reproduction:** Seeds, rhizomes.

**General Description:** An almost mythical grass because of its rarity in the Concho Valley, although historically noted as native to certain areas of this region. It has also been recorded in the Trans-Pecos region, although rarely. "Big blue" is famous for being one of the original "big four" tall bunchgrasses of the Texas prairie along with little bluestem, switchgrass, and yellow Indiangrass. This plant has been recorded growing successfully on a well-managed ranch north of San Angelo and in a small community along the roadside southeast of San Angelo, although it does appear in other locations. Other common name: turkeyfoot.

**Livestock and Wildlife Value:** Excellent and highly palatable to all classes of livestock. Typical of large bunchgrasses, this grass provides excellent nesting sites for quail.

**Management:** Since this plant is highly palatable to livestock, plant populations decrease with increased grazing pressure. Proper grazing management is critical for survival and propagation. Seeds are readily available through commercial seed sources and are sometimes included in seed mixes.

Seedhead—Bushy bluestem gets its common name from its much-branched, densely hairy seedhead. Middle Concho River. Photos by George Clendenin, USDA-NRCS.

# Bushy bluestem *(Andropogon glomeratus)*
## POACEAE—GRASS FAMILY

**Characteristics:** Native; perennial; warm-season.

**Growth Form:** Tall, robust bunchgrass; height 18 inches (leaves) to 48 inches (including seedheads).

**Reproduction:** Seeds.

**General Description:** Grows near water, especially in creek bottoms and riparian areas. The plant is easily identified by the very flat bases on the stems where they emerge from the clump. Dense, feathery seedheads form in late summer, and the fluffy seeds are blown by the wind as they mature. After frost in the fall, the entire plant turns a copper color. Elsewhere, bushy bluestem is often associated or confused with broomsedge bluestem (*A. virginicus*). As a general rule, bushy bluestem will be in lower, moist areas while broomsedge bluestem will be slightly higher on drier soils immediately above. Other common name: bushy beardgrass.

**Livestock and Wildlife Value:** Provides cattle grazing in spring and early summer; lower grazing value for sheep and goats. This grass provides little or no grazing value for white-tailed deer. No known value of seeds for birds.

**Management:** This and other riparian plants are important in helping to reduce erosion and stabilize banks and wet areas, especially during flooding. Bushy bluestem does not have the strongest roots, but it has the ability to establish readily in bare and disturbed places. Grazing in creek areas should be managed to ensure a good dense cover of riparian vegetation.

Red threeawn (var. *longiseta*) has the longest awns of all of our threeawns. Each of the 3 awns is more or less equal in length, about 3–4 inches long.

Community—Red threeawn (var. *longiseta*) is more low growing and colony forming (from seeds), and it has the longest threadlike awns of any of our other species. It is also a distinctive reddish purple in the spring and can be observed clearly along roadsides. Var. *longiseta* differs from var. *purpurea* by having longer awns and glumes and wider lemma apices and awn bases (Powell 1994). Photos by George Clendenin, USDA-NRCS.

# Red threeawn *(Aristida purpurea* var. *longiseta)*
*(Sy = A. longiseta, A. longiseta* var. *rariflora, A. longiseta* var. *robusta)*
## POACEAE—GRASS FAMILY

**Characteristics:** Native; perennial; warm-season.

**Growth Form:** Densely tufted bunchgrass; height 12–30 inches.

**Reproduction:** The 3-parted awn and sharp head can be carried by the wind or plant itself into the ground. Seeds can also be spread by sticking to passing animals.

**General Description:** In the range management profession we sometimes like to "lump and clump" similar species that may be different botanically but for all practical purposes are managed the same. Such is the case with the threeawns, which have been a source of debate and controversy among taxonomists. Recent taxonomy has merged 7 or 8 somewhat weak and controversial taxa of threeawns into one species, *Aristida purpurea.* Powell, in his work on Trans-Pecos grasses (1994), identifies 5 varieties of *A. purpurea: purpurea, longiseta, nealleyi, fendleriana,* and *wrightii.* Amos, in her complete checklist of taxa of the Concho Valley (1998), lists predominantly var. *longiseta* and var. *wrightii* as occurring in the Concho Valley. I have frequently found these same two varieties in the Concho Valley, with the possible addition of var. *purpurea.* Other common names: perennial threeawn, purple needlegrass.

**Livestock and Wildlife Value:** Threeawn provides some limited early spring grazing since it greens up earlier than most other warm-season grasses.

**Management:** This plant may increase with heavy grazing, as it replaces other, more desirable species. Ranges that are dominated by this grass indicate past overgrazing. In this condition, a comprehensive program of range management is needed to help restore a more desirable grass cover.

**Awns**—*Aristida purpurea* var. *wrightii* has shorter awns than red threeawn (var. *longiseta*). Each of the 3-parted awns is more or less equal in length, about ¾–1½ inches long.

**The "Wright" Identification**—Wright threeawn is an erect plant with bushy, curling basal leaves. The awns are shorter than those of red threeawn (var. *longiseta*), and its tan to brown (fading to straw-colored) seedhead can be distinguished from the purplish seedhead of purple threeawn (var. *purpurea*). Photos by George Clendenin, USDA-NRCS.

# Wright threeawn *(Aristida purpurea* var. *wrightii)*
*(Sy = A. wrightii)*
**POACEAE—GRASS FAMILY**

**Characteristics:** Native; perennial; warm-season.

**Growth Form:** Densely tufted bunchgrass; height 12–30 inches. Tallest of our threeawns (as compared to oldfield or red).

**Reproduction:** The 3-parted awn and sharp head can be carried by the wind or plant itself into the ground. Seeds can also be spread by sticking to passing animals.

**General Description:** Common on rocky hills and shallow soils in the region. The spikelet has 3 spreading awns, 2 of which bend out horizontally. Wright does not branch at the base and has more basal leaves than red threeawn (*Aristida purpurea* var. *longiseta*). Other common name: Wright tripleawn grass, needlegrass.

**Livestock and Wildlife Value:** Threeawn provides some limited early spring grazing since it greens up earlier than most other warm-season grasses. Threeawn seeds are encased inside a hard, sharp-pointed needle. These can penetrate the skin of animals, especially lambs, and can actually work their way into the flesh in some cases. The sharp seeds can also injure the eyes of livestock. The seeds have been known to lower the quality of wool and mohair by compacting against the animal's skin.

**Management:** This plant increases with heavy grazing, as it replaces other, more desirable species.

**Seedhead**—This seedhead has many different "looks." Here it appears mature, with the feathery seeds still remaining. Earlier, this seedhead would have appeared contracted and greenish, with the hairs all adhering tightly to the stalk. Later, after these seeds disperse, the seedhead, like those of many other grasses, will appear straw colored.

**Community**—Silver bluestem can grow in large communities and is recognizable from the road by its silvery-white inflorescence (seedhead). It is a "transitional" grass, meaning that its appearance is a sign that rangeland health may be improving. Photos taken in Christoval, Texas, by George Clendenin, USDA-NRCS.

# Silver bluestem *(Bothriochloa laguroides)*
*(Sy = B. longipaniculata)*
**POACEAE—GRASS FAMILY**

**Characteristics:** Native; perennial; warm-season.

**Growth Form:** Bunchgrass; height 15 inches (leaves) to 30 inches (including seedhead).

**Reproduction:** Seeds.

**General Description:** Coarse grass with a high ratio of stems to leaves. Lower stems often bend at the nodes instead of growing upright. Seedhead is a cluster of silvery-white, feathery seeds. As the seedhead matures, the seeds are distributed by wind. This grass is similar to and sometimes confused with the larger and more desirable cane bluestem. Silver bluestem will not have a ring of conspicuous hairs at the culm nodes (joints), or if it does the hairs will be really small, while cane bluestem will have this ring of short, spreading hairs. Cane bluestem is more widely distributed in the Trans-Pecos. Other common name: silver beardgrass.

**Livestock and Wildlife Value:** Silver bluestem is marginally palatable to cattle, sheep, and goats for a short period in spring, when new growth is tender and leafy. Afterward, grazing value is diminished and livestock tend to shun this coarse grass. It sometimes undergoes a second brief period of grazing in the fall. This grass provides little or no grazing value for white-tailed deer. Large, lightly grazed clumps provide nesting cover for quail and other ground-nesting birds.

**Management:** This is one of the grasses that responds positively to improved grazing management after decades of overgrazing. This midgrass will improve soil by adding litter and organic matter to it, thereby promoting further range improvement.

**Community**—Sideoats is extremely variable in terms of shape, ecology, and chromosome number (Diggs, Lipscomb, and O'Kennon 1999). Close observation of this healthy patch revealed no rhizomes, so it is possible this may be var. *caespitosa*. However, elsewhere on this ranch a nice patch of low-growing, thick, rhizomatous sideoats grama (var. *curtipendula*) was also found. The rhizomatous variety is very apparent in late fall and winter as the grass turns a deep purple. Photos by George Clendenin, USDA-NRCS.

# Sideoats grama *(Bouteloua curtipendula)*
## POACEAE—GRASS FAMILY

**Characteristics:** Native; perennial; warm-season.

**Growth Form:** Bunchgrass, also forms colonies from rhizomes.

**Reproduction:** Seeds and rhizomes, depending on variety.

**General Description:** The State Grass of Texas as adopted by the 62nd Texas legislature in 1971. This grass is one of the more desirable species in the Concho Valley, for both livestock operators and wildlife managers alike. All *Bouteloua* species are warm-season plants with a leaf anatomy that translates to more effective capture of carbon dioxide and thus less water loss through transpiration (Diggs, Lipscomb, and O'Kennon 1999). This is always a good thing in a semiarid environment like that of the Concho Valley. Sideoats grama is a midgrass with a solid root structure that is effective at increasing ground infiltration of water at the ranch level, and it is highly palatable to livestock. As with other bunchgrasses, it provides nesting sites for quail. There are two native varieties of *B. curtipendula,* one with rhizomes (var. *curtipendula*), and one without rhizomes (var. *caespitosa*). The rhizomatous form is much more palatable and preferred over the bunch form, but the bunch form is more productive.

**Livestock and Wildlife Value:** Sideoats grama is a desirable plant for livestock and wildlife. Its forage is palatable to all classes of livestock during the growing season and it provides nesting cover for quail and other grassland birds. Turkeys eat the seeds by "stripping" the seedheads.

**Management:** Since sideoats grama is palatable, plant densities will decrease with increased grazing pressure. Proper grazing management is critical for survival and propagation. The NRCS James E. "Bud" Smith Plant Materials Center in Knox City, Texas, released a southern variety called 'Haskell,' which was selected for its forage productivity and rhizomatous production. 'Haskell' is frequently used in range plant seed mixtures and performs well in west central Texas.

**Native American Uses:** The seedheads of sideoats grama were worn as decoration by Kiowa warriors who killed an enemy with a lance in battle (because the seedheads resemble a feathered lance).

Seedhead—Hairy grama is characterized by the "stinger" at the end of each branched seedhead. It doesn't sting nor is it sharp, but it does extend past the spikelets and appear pointy. Photos by George Clendenin, USDA-NRCS.

# Hairy grama *(Bouteloua hirsuta)*
## POACEAE—GRASS FAMILY

**Characteristics:** Native; perennial; warm-season.

**Growth Form:** Short bunchgrass; height 6 inches (leaves) to 12 inches (including seedhead).

**Reproduction:** Seeds.

**General Description:** Leaves are thin, short, and hairy upon close examination. Seedhead consists of 1–4 short branches per stem, with each branch composed of many small seed-bearing units arranged in rows like the teeth in a comb. Seedhead branches are hairy and terminate in a pointed tip. Usually grows in shallow, rocky soil.

**Livestock and Wildlife Value:** Provides some grazing for livestock while green and growing, but not typically preferred forage. Because of its shallow root system, this grass does not stay green as long as taller grasses. Hairy grama provides little or no wildlife value. It does not provide nesting cover, the seeds have no known value for birds, and it provides very little grazing value for white-tailed deer.

**Management:** This plant is not considered desirable by ranchers or wildlife managers. The value of hairy grama is to provide ground cover and rooting to help hold fragile soils in place. With good range management, this grass will often slowly decline and be replaced by taller and deeper-rooted grasses such as sideoats grama.

**Seedhead**—The seedhead of Texas grama vaguely resembles that of sideoats grama to the untrained eye, but it is much shorter and has many fewer branches. Photos by George Clendenin, USDA-NRCS.

# Texas grama *(Bouteloua rigidiseta)*
## POACEAE—GRASS FAMILY

**Characteristics:** Native; perennial; warm-season.

**Growth Form:** Short bunchgrass; height 6 inches (leaves) to 10 inches (including seedhead).

**Reproduction:** Seeds.

**General Description:** This short grass can be difficult to identify without the seedheads. Seedheads are composed of 5–12 short, wedge-shaped branches off the central stalk. Each branch contains chaffy material and several seeds. Once the seeds have shattered and the naked stalk remains, a keen observer will still be able to identify this grass.

**Livestock and Wildlife Value:** Texas grama provides very limited grazing for livestock and is not considered desirable forage. This grass provides no grazing value for white-tailed deer, seeds are of no known value for birds, and the plant is too small to provide nesting cover.

**Management:** This short grass is characteristic on pastures that have been heavily grazed in the past. It can provide good ground cover and help hold soil in place until other, more desirable plants become established. With good grazing management, this plant will be gradually replaced by taller, more desirable grasses.

Photos by George Clendenin, USDA-NRCS.

# Red grama *(Bouteloua trifida)*
## POACEAE—GRASS FAMILY

**Characteristics:** Native; perennial; warm-season.

**Growth Form:** Very short bunchgrass; height 3 inches (leaves) to 8 inches (including seedhead).

**Reproduction:** Seeds.

**General Description:** This plant has the smallest leaves of any grass in the Concho Valley, from ¾ to 2 inches long. Seedheads consist of 2–7 seed-bearing branches. Each branch contains many seed units enclosed in chaffy material with awns. Seedheads are reddish or pinkish overall but lose color after maturity.

**Livestock and Wildlife Value:** Provides poor grazing for cattle because of small plant size and limited ability of cattle to obtain bites. Red grama provides limited grazing to sheep and goats while green and growing, but its season of growth is brief because of its shallow root system. Red grama provides no food value for white-tailed deer or birds.

**Management:** The best that can be said for red grama is that it helps hold soil in place even on pastures that have been heavily grazed for decades. As grazing management is improved, this grass will gradually be replaced by taller, more deeply rooted plants that provide better grazing and better wildlife habitat. Pastures dominated by red grama are in a state of very poor health and have possibly had severe erosion in the past. Drastic changes in grazing management and/or reseeding are often needed to begin range restoration.

Seedhead—Spikelets are much flattened, and without hairs.

Community—Although now considered more of a weed, rescue grass got its common name from coming to the "rescue" of stockmen by being the first green forage after drought or a long winter (Hatch and Pluhar 1993). Photos by George Clendenin, USDA-NRCS.

# Rescue grass *(Bromus catharticus)* *(Sy = B. brevis, B. haenkeanus, B. unioloides, B. willdenowii)*
## POACEAE—GRASS FAMILY

**Characteristics:** Introduced; annual; cool-season.

**Growth Form:** Bunchgrass; height 12 inches (leaves) to 24 inches (including seedhead).

**Reproduction:** Seeds.

**General Description:** This large leafy grass is native to South America and was introduced as a forage grass. It is a common yard weed and is also found on many ranches. It is sometimes incorrectly referred to as "winter rye." Plants germinate in late fall through winter when moisture conditions are good. Rescue grass grows slowly until late winter and then begins to grow rapidly. By early spring plants are large and seedheads are formed.

**Livestock and Wildlife Value:** Good grazing for livestock during late winter and early spring when plants are leafy, also grazed by white-tailed deer during this time. As plants mature in midspring, grazing value quickly declines. Tender leaves are grazed by turkeys. Green seedheads are stripped by turkeys, and quail occasionally eat the shattered seeds.

**Management:** A good cover of native grass will reduce or prevent the growth of cool-season annual grasses. A dense cover of rescue grass often indicates an abundance of bare ground and a poor cover of native grasses. Although the grazing is good, it is short-lived and unreliable, providing substantial forage only in wet years. A disadvantage of rescue grass is that it often robs soil moisture from warm-season grasses and retards their spring and early summer growth.

Seedhead—Male or Female? Buffalograss grows as separate male and female plants. Male plants have elevated seedheads that produce pollen. Female plants produce no true seedhead, but the female flower and seed are located in a bur-type structure near the base of the plant. The seedheads pictured are on a male plant. Photos by George Clendenin, USDA-NRCS.

# Buffalograss *(Buchloe dactyloides)*
## POACEAE—GRASS FAMILY

**Characteristics:** Native; perennial; warm-season.

**Growth Form:** Sod-forming grass that grows in large colonies; height 6 inches (leaves) to 10 inches (including seedhead).

**Reproduction:** Seeds and stolons (runners).

**General Description:** One of the two main sod-forming native grasses in the region and sometimes confused with curly-mesquite grass. Buffalograss stolons are completely smooth to the touch, while the stolons of curly-mesquite are hairy and rough to the touch. Buffalograss grows best in deep clay or clay loam soils and is not commonly found in shallow, rocky soil.

**Livestock and Wildlife Value:** Provides excellent grazing for all classes of livestock. This grass cures well for winter grazing. Although it is one of the most palatable and nutritious grasses, its volume of forage is considerably lower than that of taller grasses. It offers limited forage value for white-tailed deer. The seeds are sometimes eaten by various species of grassland sparrows.

**Management:** This grass will often be heavily grazed unless rest periods are provided where livestock are temporarily moved to other pastures. Buffalograss can spread rapidly by runners under the right conditions. Careful attention to grazing management will allow buffalograss to increase. Several varieties of buffalograss are commercially available for range seeding and for turf and lawn establishment. Buffalograss lawns require no fertilizer, little or no mowing, and very little watering.

Seedhead—Hooded windmillgrass and tumble windmillgrass share the radiating branched panicle. Both grasses can have between 10 and 20 panicle branches. Hooded windmillgrass, however, is the smaller of the two; its seedhead is only 1½–2½ inches across. Photos by George Clendenin, USDA-NRCS.

# Hooded windmillgrass *(Chloris cucullata)*
## POACEAE — GRASS FAMILY

**Characteristics:** Native; perennial; warm-season.

**Growth Form:** Bunchgrass; height 8 inches (leaves) to 16 inches (including seedhead).

**Reproduction:** Seeds.

**General Description:** This grass is more commonly seen in loam and sandy loam soils. It is less commonly found in clay loam soils. The basal stems are very strongly flattened, and these are the best way to identify the grass without a seedhead. Seedhead branches all emerge from a central location and resemble the blades of a windmill.

**Livestock and Wildlife Value:** Provides good grazing for livestock and stays green longer than some of the short grasses. This grass provides little or no grazing value for white-tailed deer and there is no documented seed value for birds.

**Management:** This grass is best maintained by proper grazing management, which consists of using moderate stocking rates, adjusting livestock numbers during drought, and providing intermittent rest periods.

Seedhead—Tumble windmillgrass has a much larger seedhead, with a diameter of 4–8 inches, than its cousin, hooded windmillgrass. Photos by George Clendenin, USDA-NRCS.

# Tumble windmillgrass *(Chloris verticillata)*
## POACEAE—GRASS FAMILY

**Characteristics:** Native; perennial; warm-season.

**Growth Form:** Primarily a bunchgrass, yet also forms mats because of rooting of stems; height 6 inches (leaves) to 16 inches (including seedhead).

**Reproduction:** Seeds and rooting at nodes. After the seedhead breaks away from the plant, it is blown by the wind and tumbles across the ground distributing seed.

**General Description:** This plant is seen more often in disturbed locations and not as frequently on native rangeland. Tumble windmillgrass closely resembles hooded windmillgrass, with both having flattened stem bases. This grass crosses with hooded windmillgrass and produces hybrid plants that may be intermediate in appearance.

**Livestock and Wildlife Value:** Grazed by livestock but not preferred. This grass provides little or no grazing value for white-tailed deer and has no known use by birds.

**Management:** This grass requires no specific management.

**Leaf ID Tip**—Positive identification can be attained by examining the leaves, which have a fine crinkled edge on one side, resembling a hacksaw blade.
Photos by George Clendenin, USDA-NRCS.

# Fall witchgrass *(Digitaria cognata)*
POACEAE—GRASS FAMILY

**Characteristics:** Native; perennial; warm-season.

**Growth Form:** Bunchgrass; height about 12 inches including seedhead.

**Reproduction:** Seeds.

**General Description:** Bushy grass with weak stems, often seen growing in shrubs and cactus for support and protection. Seedhead resembles a small broom that is compact when new and spread out when mature. Mature seedheads often break away from the plant and tumble across the ground to spread the seeds.

**Livestock and Wildlife Value:** Readily grazed by livestock. This grass provides little or no grazing value for white-tailed deer. Dense, large clumps provide some nesting cover for ground-nesting birds. Use of seeds by birds is presumed but not documented.

**Management:** This grass responds positively to good grazing management.

Seedhead ID Tip—Although Virginia wildrye is not common in the area, there are ways to distinguish it from Canada wildrye: the glumes at the base of the Canada wildrye seedhead form a *V*, whereas those of Virginia wildrye form a *U*. Next, the seedhead of Canada wildrye will droop at maturity, whereas the seedhead of Virginia wildrye remains erect. The leaves of Canada wildrye are also wider than those of Virginia.

Community—This photo was taken in May after a spring of favorable precipitation; therefore the Canada wildrye is still green and vigorous. The seedhead will droop at maturity and turn a straw yellow. Photos by George Clendenin, USDA-NRCS.

# Canada wildrye *(Elymus canadensis)*
## POACEAE—GRASS FAMILY

**Characteristics:** Native; perennial; cool-season.

**Growth Form:** Bunchgrass; height 15 inches (leaves) to 30 inches (including seedhead).

**Reproduction:** Seeds.

**General Description:** Tall, leafy grass with seedheads that vaguely resemble those of wheat or cereal rye. This grass grows on a wide variety of soils from creek bottoms to deep clay loam flats to rocky hillsides. Virginia wildrye also occurs in the region but is much less common.

**Livestock and Wildlife Value:** Excellent grazing for livestock during late winter and early spring when green and growing. Some use by white-tailed deer when leaves are young and tender. Large old clumps may be used by turkeys to help conceal nests.

**Management:** Careful attention to grazing management is required to maintain or increase this grass. For those who wish to enhance the cool-season forage production of pastures, seeds of this grass are commercially available and can be added to seed mixtures. The NRCS E. "Kika" de la Garza Plant Materials Center in Kingsville, Texas, released a southern variety called 'Lavaca Germplasm,' which was selected for its forage productivity. 'Lavaca Germplasm' can be used as a cool-season component of a native seed mixture for range restoration or as a cool-season pasture.

Photos by
George Clendenin,
USDA-NRCS.

# Squirreltail *(Elymus longifolius)* (Sy = *E. elymoides, Sitanion hystrix*)
**POACEAE—GRASS FAMILY**

**Characteristics:** Native; perennial; cool-season.

**Growth Form:** Bunchgrass; height 12 inches (leaves) to 18 inches (including seedhead).

**Reproduction:** Seeds.

**General Description:** Squirreltail prefers growing in mottled shade and can sometimes be found under a motte of live oak trees. The plant can appear odd because it grows by itself and has seedheads with abundant, extremely long awns (hairlike structures). This is a cool-season plant, so it does not favor our hot, droughty summers and our semiarid climate, but surprisingly, it is recorded as perhaps the most common *Elymus* in the Trans-Pecos (Powell 1994). Other common name: longleaf squirreltail.

**Livestock and Wildlife Value:** Provides grazing for livestock during spring. This grass provides little or no grazing value for white-tailed deer and has no known use by birds.

**Management:** This grass is usually present in very small amounts, making it difficult to manage. Squirreltail can be maintained by good grazing management practices.

Seedhead—The seedhead of plains lovegrass is oval in outline. The spikelets are very small, but each one can contain 5–11 florets (as opposed to a similar-appearing species of *Muhlenbergia,* also known as muhly grass, which usually has only 1 floret per spikelet). Photos by George Clendenin, USDA-NRCS.

# Plains lovegrass *(Eragrostis intermedia)*
POACEAE—GRASS FAMILY

**Characteristics:** Native; perennial; warm-season.

**Growth Form:** Bunchgrass; height 12 inches (leaves) to 24 inches (including seedhead).

**Reproduction:** Seeds.

**General Description:** A very leafy bunchgrass that greens up earlier in spring than most other warm-season grasses. Seedheads are open and much branched and often partially enclosed in the leaf sheath. The seedheads often break loose from the plant in the fall and are blown across the ground, sometimes accumulating along fences. Plains lovegrass is the only *Eragrostis* (lovegrass) listed in this manual; however, Amos (1998) also recorded seven other *Eragrostis* species in the valley: Mediterranean, stinkgrass, gummy, Lehman, red, tumble, and sand lovegrasses.

**Livestock and Wildlife Value:** Provides very good grazing value for livestock during growing season. This grass provides little or no grazing value for deer. Large clumps of plains lovegrass that are grazed lightly or protected by cactus provide good nesting cover for quail and other ground-nesting birds. The seeds are too tiny to be of value to birds.

**Management:** This desirable forage grass is underappreciated and can be enhanced by good conservative grazing management, including rotational grazing.

Seedheads are distinctive, with several vertical rows of seeds held tightly to the central stalk. Individual spikelets appear to sit in a cuplike chair; thus the common name "cupgrass."

A large community of Texas cupgrass along the bank of Pecan Creek. Photos by George Clendenin, USDA-NRCS.

# Texas cupgrass *(Eriochloa sericea)*
## POACEAE—GRASS FAMILY

**Characteristics:** Native; perennial; warm-season.

**Growth Form:** Bunchgrass; height 15 inches (leaves) to 30 inches (including seedhead).

**Reproduction:** Seeds.

**General Description:** Cupgrass is usually found growing in shallow, rocky soil and less often in deep soil. It is one of the larger bunchgrasses of the region. Seedheads may form 2 or 3 times during the growing season in response to rainfall. Other common name: silky cupgrass.

**Livestock and Wildlife Value:** Very good grazing value for livestock. This grass provides little or no grazing value for deer. Turkeys, quail, and other grassland birds eat Texas cupgrass seeds. Large clumps of Texas cupgrass provide good nesting cover for ground-nesting birds.

**Management:** This very desirable forage grass can be maintained only by the conscientious practice of good progressive grazing management. Under these conditions, the grass will increase. In situations of overgrazing, plant densities will decrease. This and many other desirable forage grasses often find refuge within clumps of prickly pear or other spiny shrubs. These protective areas are often important for the survival and eventual increase of these grasses.

Seedhead—The seedhead on the left is still green with reddish-tipped spikelets, while the seedhead on the right is more mature and dried out and beginning to express the typical hairs. The common names "hairy woollygrass" and "hairy erioneuron" are derived from these hairy mature seedheads.

Grass ID Tip—In the field, hairy tridens can be identified by certain clues: it has short, blunt-tipped leaves with no hair; a fine white margin on the leaves; and leaf sheaths that are "keeled," meaning shaped like the bottom of a ship. Prescribed burning—This plant was photographed in September, 7 months after the entire pasture was burned with a winter burn in February. It demonstrates how vigorous and healthy our warm-season perennial plants can be when they come back after a prescribed burn. Photos by George Clendenin, USDA-NRCS.

# Hairy tridens *(Erioneuron pilosum)*
## POACEAE—GRASS FAMILY

**Characteristics:** Native; perennial; warm-season.

**Growth Form:** Short bunchgrass; height 2 inches (leaves) to 8 inches (including seedhead).

**Reproduction:** Seeds.

**General Description:** This is one of the shortest and least productive of all grasses in the region. Mature seedheads are full of dense hair upon close examination, but the hair is less visible when the seedheads are closed. Other common names: hairy woollygrass, hairy erioneuron.

**Livestock and Wildlife Value:** Because of its very short stature, this plant has minimal forage value to cattle. It is grazed to a limited extent by sheep and goats but provides little or no grazing value for white-tailed deer.

**Management:** Pastures that are dominated by a large percentage of hairy tridens indicate a past history of overgrazing. Heavily grazed pastures with poor grass cover can be restored by the application of progressive range management practices including proper stocking rates, frequent long rest periods, or reseeding.

**Stolons**—Compared to the stolons of buffalograss, those of curly-mesquite are longer and more slender and are covered by hairs, giving a slightly rough texture. The stolons of buffalograss are shorter, thicker, and smooth to the touch. Some grass experts tell the difference by running the stolons across their lip, where they can more readily feel the rough or smooth texture. Photos by George Clendenin, USDA-NRCS.

# Curly-mesquite *(Hilaria belangeri)*
## POACEAE—GRASS FAMILY

**Characteristics:** Native; perennial; warm-season.

**Growth Form:** Sod-forming grass; height 6 inches (leaves) to 10 inches (including seedhead).

**Reproduction:** Seeds and stolons (runners).

**General Description:** One of the two main sod-forming grasses in the region and often confused with buffalograss. Curly-mesquite grows more commonly in rocky and shallow soil but can also sometimes be found in deeper clay loam soils.

**Livestock and Wildlife Value:** Provides fair grazing for livestock but is much inferior to buffalograss. This plant has little or no use by white-tailed deer and no known use by birds.

**Management:** This grass is important for its ability to spread rapidly by runners. Under the right conditions it can colonize bare ground in a short time and fill the gaps between larger bunchgrasses. Its grazing value is probably less important than its value as ground cover and soil protection. This grass needs no specialized management except to avoid extreme chronic overgrazing.

Prescribed fire—This plant was photographed in September, 7 months after the entire pasture was burned with a winter burn back in February. It demonstrates how vigorously desirable plants come back after a prescribed burn. It also shows how important it is to rest pastures completely from livestock for a full growing season immediately after a winter burn. Photos by George Clendenin, USDA-NRCS.

# Green sprangletop *(Leptochloa dubia)*
## POACEAE—GRASS FAMILY

**Characteristics:** Native; perennial; warm-season.

**Growth Form:** Bunchgrass; normal height 15 inches (leaves) to 30 inches (including seedhead).

**Reproduction:** Seeds.

**General Description:** This grass is found naturally on shallow, rocky soils on hillsides and in canyons, but it will grow in nearly any soil type if planted. Green sprangletop is cyclic in abundance and has difficulty competing with the hardiness of other perennial grasses. When conditions are right, it can establish quickly and be abundant for a few years, after which it declines. Abundant seeds are produced and lie in the soil waiting for the proper conditions.

**Livestock and Wildlife Value:** Excellent livestock grazing during the growing season. Limited value as white-tailed deer forage. Provides nesting cover for ground-nesting birds.

**Management:** In order for this short-lived grass to perpetuate itself, it must be able to make good seed crops. Proper grazing management will allow plants to maintain vigor and produce seeds. This grass is widely planted as a minor component of range seeding mixtures. It establishes more quickly than most other grasses, providing a fast initial cover. This and other desirable forage grasses are best maintained by the careful practice of good conservative grazing management. This grass responds positively to fire with a flush of new growth.

Seedhead—Three-flower melic's seedhead faintly resembles that of domestic rice. Photo by George Clendenin, USDA-NRCS.

Photo by Steve Nelle, USDA-NRCS.

# Three-flower melic *(Melica nitens)*
## POACEAE—GRASS FAMILY

**Characteristics:** Native; perennial; cool-season.

**Growth Form:** Bunchgrass; height 15 inches (leaves) to 30 inches (including seedhead).

**Reproduction:** Seeds.

**General Description:** This desirable cool-season, large, leafy grass is most often found in rocky or shallow soils in canyons and headers, especially in protected locations, but it will also grow in deep soil. Plant begins to green up in midwinter and grows rapidly in March and April when most other grasses are dormant. Seedheads begin to form in April and are mature in May.

**Livestock and Wildlife Value:** Excellent grazing value in late winter and early spring. Probably grazed to some extent by white-tailed deer, but this is undocumented. Turkeys eat the seeds by "stripping" the seedheads. Seeds are large enough to be valuable to birds, but seed consumption is unconfirmed.

**Management:** This desirable and palatable cool-season grass is difficult to manage properly because it makes up only a small proportion of total forage production. It is often found growing only in the protection of catclaw or other shrubs. This grass can be expected to increase only with the careful application of conservative rotational grazing. It is under consideration for commercial seed production.

**Mature Seedhead**—After the seedheads (spears) fall off, papery white bracts (glumes) remain. In absence of seedheads, however, the best way to identify this grass is to run the leaf blade through your fingers to see whether you can feel the short bristly "hairs."

**Spears Still Remain**—This plant still has its spears (spikelets), providing ready ammunition for spear-throwing kids (and adults)! Photos by George Clendenin, USDA-NRCS.

# Texas wintergrass *(Nassella leucotricha)*
*(Sy = Stipa leucotricha)*
**POACEAE — GRASS FAMILY**

**Characteristics:** Native; perennial; cool-season.

**Growth Form:** Bunchgrass; height 12 inches (leaves) to 18 inches (including seedhead).

**Reproduction:** Seeds.

**General Description:** Commonly known as "speargrass," this grass produces a sharp-pointed seed and an awn that resembles a spear. The leaves are covered by very short, stiff hairs that give them a slightly rough texture. Often grows in association with mesquite and is most common in deep clay loam soils. Other common name: speargrass, Texas needlegrass.

**Livestock and Wildlife Value:** Excellent grazing for all classes of livestock in late winter and spring. As the plant matures, grazing value diminishes. Wintergrass provides some late winter and early spring grazing for white-tailed deer and turkeys. In some cases, wintergrass may grow too dense for quail and impede their movement. Where it grows in clumps with spaces between, quail sometimes use it as nesting cover. Wool sheep, especially lambs, may be injured if seeds get lodged in wool and penetrate the skin.

**Management:** This is often the primary forage grass for livestock in the winter and early spring. To maintain good wintergrass pastures, graze only during a portion of the winter-spring season and provide rest during mid to late spring. Early to midwinter burning may be used to increase the nutritional quality of forage.

**Wood Shavings**—This is what the curled, dried leaves look like when they are mature, providing a useful plant ID tool in the field. Photos by George Clendenin, USDA-NRCS.

# Hall panicum *(Panicum hallii)*
## POACEAE—GRASS FAMILY

**Characteristics:** Native; perennial; warm-season.

**Growth Form:** Short bunchgrass; height 6 inches (leaves) to 12 inches (including seedhead).

**Reproduction:** Seeds.

**General Description:** This grass is easily identified late in the growing season by the unique curling of the leaves, which resemble wood shavings. This characteristic is not found on new green leaves, but only old dry leaves. Panicles are open and spikelets contain hard, shiny seeds. It is most commonly found on barren or sparsely vegetated areas. Other common names: Hall's panic grass; Hall's panicum.

**Livestock and Wildlife Value:** Hall panicum provides some limited grazing for livestock but it is not preferred. This grass provides little or no grazing value for white-tailed deer. Quail and other ground birds occasionally eat the seeds.

**Management:** This shallow-rooted grass grows only when the soil is moist near the surface. This grass can establish well on bare ground following drought when conditions are right. Although it does not provide much desirable forage, it does provide cover on the soil and a root system that helps hold soil in place. Such "pioneer plants" help to begin the process of land healing. A pasture with a significant cover of Hall panicum generally indicates a chronic history of heavy grazing.

Seedhead—Seedheads are composed of several rows of large plump seeds that stay close to the central stalk.

Community—The darker green and reddish patch shown here is a community of vine-mesquite. This is easily distinguishable from the adjacent vegetation and may indicate a low spot where water may accumulate. Photo taken at a cattle ranch in Christoval, Texas. Photos by George Clendenin, USDA-NRCS.

# Vine-mesquite *(Panicum obtusum)*
## POACEAE—GRASS FAMILY

**Characteristics:** Native; perennial; warm-season.

**Growth Form:** Midgrass that forms large patches; height 12 inches (leaves) to 16 inches (including seedhead).

**Reproduction:** This is one of the few grasses that spread by three methods—seeds, stolons (runners), and rhizomes.

**General Description:** This grass is most often found on deeper soils and low spots where extra water accumulates, but it is also sometimes seen on flat-topped hills and divides. The most characteristic trait of this plant is its long, trailing stolons, often called runners, which crawl across the ground and have the ability to root at each node and make new plants. These runners are often 4–8 feet long, showing the tremendous ability of this plant to spread.

**Livestock and Wildlife Value:** Vine-mesquite provides good livestock grazing when tender and green, but palatability declines when the grass becomes dry or dormant. One local rancher has dubbed this grass "cow spaghetti" because of his observations that cattle readily eat this grass, and with the long runners hanging out of their mouths they look like kids eating noodles. This grass provides little or no grazing value for white-tailed deer. Patches of this grass provide good nesting cover for quail and other grassland birds. The seeds have been reported to be valuable for quail and doves. However, most of the seeds are sterile and have no actual kernel; they appear plump but are mostly empty on the inside.

Community—Switchgrass was successfully used here as part of the NRCS grass seed mixture to prevent soil erosion on an erodible water channel meandering through the cropland field. You can see the ability of this plant to produce biomass in this situation. The fibrous root system that you do not see, however, is also performing its job of stabilizing the soil.

Streambank, South Concho River. Photos by George Clendenin, USDA-NRCS.

# Switchgrass *(Panicum virgatum)*
## POACEAE — GRASS FAMILY

**Characteristics:** Native; perennial; warm-season.

**Growth Form:** Very large bunchgrass; height 3 feet (leaves) to 6 feet or more (including seedhead).

**Reproduction:** Seeds.

**General Description:** Historically, switchgrass was one of the original "big four" tall bunchgrasses of the Texas prairie along with little bluestem, big bluestem, and yellow Indiangrass. In the Concho Valley region, this grass is common in well-managed creek and river bottom areas. Switchgrass has 2 distinct growth forms: (1) lowland switchgrass, which grows in large isolated clumps and is associated with low, moist areas; and (2) upland switchgrass, which is smaller but strongly rhizomatous and is seen growing in drier, upland sites. Switchgrass is also being extensively investigated as a potential biofuel source.

**Livestock and Wildlife Value:** Switchgrass provides a very large quantity of forage for cattle (less suitable for sheep and goats), but it becomes coarse with age, which reduces palatability and nutritional quality. This grass provides little or no grazing value for white-tailed deer. The overhanging leaves provide nest concealment for turkeys. Switchgrass also provides general screening cover for deer, especially fawning cover.

**Management:** In native creek bottom areas, switchgrass is managed by controlling the timing and intensity of grazing. Short periods of grazing followed by long rest periods will allow switchgrass to grow to its fullest extent. The dense, deep, binding root system is important to help stabilize creek banks and reduce erosion during flooding. The large, dense clumps also help dissipate the energy of floodwaters and trap sediment in floodplain areas. Switchgrass seeds are commercially available and can be planted as a component of a native grass mix. The 'Alamo' variety grows as very large, robust clumps; the smaller 'Blackwell' variety is a superior forage grass that grows as a colony with rhizomes. In upland areas, the best grazing will be in spring when the leaves are tender. Burning will help reduce old growth and improve the nutritional quality of forage.

**Native American Uses:** When the Pawnees cut up meat on the buffalo hunt, they were careful to avoid laying it on top of this grass when it was in head, because the glumes of the spikelets would adhere to the meat and afterward would stick in the throat of one eating it (Gilmore 1919).

Seedhead—The tall, erect, spike-type seedhead has alternating spikelets, each overlapping the one above by half its length. Wheatgrass has a characteristic blue-green color when growing and eventually will turn straw yellow when mature and drying.

Community—These wind turbine rights-of-way were range planted with a seed drill with western wheatgrass; however, the planting of western wheatgrass on low stony hills or shallow ecological sites is not recommended. Photos by George Clendenin, USDA-NRCS.

# Western wheatgrass *(Pascopyrum smithii)*
*(Sy = Elytrigia smithii, Agropyron smithii)*
**POACEAE—GRASS FAMILY**

**Characteristics:** Native; perennial; cool-season.

**Growth Form:** Colony-forming grass; height 12 inches (leaves) to 24 inches (including seedhead).

**Reproduction:** Seeds and rhizomes.

**General Description:** This grass grows naturally in the deep bottomlands of the Concho River and major creeks but is uncommon. It can grow in full sun or under the canopy of hardwood trees. The leaves are a distinctive blue gray and the veins are very evident. Large colonies may form from the very aggressive rhizomes.

**Livestock and Wildlife Value:** Provides good grazing for livestock in early spring while still green, but forage quality declines as it goes dormant in summer. White-tailed deer utilize western wheatgrass on a limited basis, especially if green forbs are not available. Large, lightly grazed colonies provide good nesting cover for ground-nesting birds. Turkeys will eat western wheatgrass seeds by "stripping" the seedheads.

**Management:** This grass can be maintained by good conservative grazing management. Seeds are commercially available and can be added as a cool-season component of a range seeding mixture. This grass is effective in deep clay loam soils or as part of good waterway vegetation to prevent soil erosion.

Community—Knotgrass is an important stabilizer of creek banks in the Concho Valley because of its ability to root at each node and its extensive root structure. Photos by George Clendenin, USDA-NRCS.

# Knotgrass *(Paspalum distichum)*
## POACEAE—GRASS FAMILY

**Characteristics:** Native; perennial; warm-season.

**Growth Form:** Colony-forming grass; height 12 inches (leaves) to 16 inches (including seedhead).

**Reproduction:** Seeds, stolons (runners), and rhizomes.

**General Description:** This grass grows only in wet areas such as creek banks, seeps, or the edges of ponds or playa lakes. Large colonies can form, primarily by the long runners that root every few inches, forming a dense mat. Seedhead has two branches in a *V* shape, with rows of seeds on each branch.

**Livestock and Wildlife Value:** Excellent grazing for livestock. Since it grows in wet areas, knotgrass tends to stay green longer than upland grasses. There is some limited use by white-tailed deer when the grass is young and tender. Turkeys and waterfowl consume the seedheads and seeds.

**Management:** This grass is important in riparian areas, where its primary function is to protect banks, stabilize soil, and reduce erosion. Research indicates that knotgrass can produce 18 miles of roots in 1 cubic foot of soil, providing exceptional binding and reinforcement. Grazing in riparian areas should be light and intermittent to maintain good cover and protection of banks.

Seedhead—Tobosagrass can be confused with curly-mesquite. They are both in the same genus (*Hilaria*). The spikelets on tobosagrass seedheads are more fan shaped and appear papery with torn edges.

Community—Tobosagrass is sometimes seen growing as an individual clump but often grows as large, dense colonies interconnected by rhizomes. Photos by George Clendenin, USDA-NRCS.

# Tobosagrass *(Pleuraphis mutica) (Sy = Hilaria mutica)*
## POACEAE—GRASS FAMILY

**Characteristics:** Native; perennial; warm-season.

**Growth Form:** Midgrass forming dense colonies; height about 12 inches (leaves) to 16 inches (including seedhead).

**Reproduction:** Seeds and rhizomes.

**General Description:** Leaves are coarse and tough compared to those of other grasses. Tobosagrass is slow to decompose because of the high lignin content in its leaves and stems. Often, three years' worth of grass can be observed, with the current year's growth being green, the previous year's growth being brown, and the two-year-old growth being gray. The largest areas of tobosagrass grow on deep clay flat soils as found on broad divides; also found on deep clay loam soils, but not frequently seen on rocky or shallow soils. Other common name: tobosa.

**Livestock and Wildlife Value:** Although this grass is notoriously tough and coarse when mature, the young spring growth provides good grazing for livestock. This grass provides little or no grazing value for white-tailed deer. The large clumps provide good nesting cover for quail and other grassland birds.

**Management:** Abundant tobosagrass can be an asset to ranchers if properly managed. Tobosagrass flats can be burned in late winter or early spring, when there is good soil moisture. New tender spring growth provides a great abundance of high-quality grazing for a few months.

**Fall Color**—At frost, little bluestem often turns its signature copper brown color.

**Community**—This little bluestem community was photographed in October, 8 months after the entire pasture was burned with a prescribed winter burn back in February. It demonstrates how vigorously desirable plants come back after a prescribed burn. It also shows how important it is to rest pastures completely from livestock for a full growing season immediately after a winter burn. Photo taken at cattle ranch near Christoval, Texas. Photos by George Clendenin, USDA-NRCS.

# Little bluestem *(Schizachyrium scoparium)*
**POACEAE — GRASS FAMILY**

**Characteristics:** Native; perennial; warm-season.

**Growth Form:** Bunchgrass; height 15 inches (leaves) to 30 inches (including seedhead).

**Reproduction:** Seeds.

**General Description:** In this region, little bluestem grows primarily in shallow, rocky soils. In other regions it grows in deep sandy or clay soils. Individual stems of this plant are very flattened at the base, which helps with identification when seedheads are not present. Seedheads begin to elongate in late summer and mature in fall, producing only a single seed crop each year.

**Livestock and Wildlife Value:** Good livestock grazing in spring and early summer when new growth is tender. Becomes tough and unpalatable by midsummer through frost. This grass does not cure well and does not provide good winter forage. It provides little or no grazing value for white-tailed deer. The overhanging leaves of lightly grazed plants provide excellent nesting cover for quail and other grassland birds.

**Management:** Formerly very common across the Concho Valley, this grass has been reduced in abundance because of improper grazing management, especially in the early 1900s. Because of improved grazing practices and lower livestock numbers, this grass is now increasing on well-managed ranches. It responds very positively to winter burning. The value of this grass can be easily misinterpreted. Since it is not grazed much in summer or fall, the large clumps with seedheads sometimes appear to be unpalatable; however, it provides a large volume of good grazing early in the growing season.

Seedhead—This bristly seedhead of plains bristlegrass has a cylindrical shape and is not tapered. Its close cousin, southwestern bristlegrass (*S. scheelei*), has a tapered seedhead, mostly longer bristles, and broader leaves.

Community—Plains bristlegrass is seen vividly among the native grass community. Photos by George Clendenin, USDA-NRCS.

# Plains bristlegrass *(Setaria leucopila)*
## POACEAE—GRASS FAMILY

**Characteristics:** Native; perennial; warm-season.

**Growth Form:** Bunchgrass; height 15 inches (leaves) to 30 inches (including seedhead).

**Reproduction:** Seeds.

**General Description:** Plains bristlegrass prefers open, upland sites. It grows more commonly in the western half of the region, especially in loamy or sandy loam soils; it is not frequently found in clay loam or shallow soils. Seedheads are dense spikes of seeds with many bristled awns. It is widely distributed and frequent in all counties of the Trans-Pecos region (Powell 1994). A similar grass, southwestern bristlegrass (*S. scheelei*), seems to prefer the shade.

**Livestock and Wildlife Value:** Grazed well by livestock during the growing season and perhaps some in winter. This grass provides little or no grazing value for white-tailed deer. Seeds are eaten by quail and other grassland birds. Larger clumps of grass provide nesting cover for quail.

**Management:** This grass responds positively to good grazing management. Seeds are commercially available for addition to range seeding mixes in loamy or sandy loam soils.

Seedhead—Close examination will reveal very short, bristly hairs originating just below the spikelets and extending just past them. Photos by George Clendenin, USDA-NRCS.

# Reverchon bristlegrass *(Setaria reverchonii)*
## POACEAE—GRASS FAMILY

**Characteristics:** Native; perennial; warm-season.

**Growth Form:** Small bunchgrass; height 8 inches (leaves) to 16 inches (including seedhead).

**Reproduction:** Seeds.

**General Description:** Found on very shallow soils. Seedheads are a thin spike with few seeds. Reverchon resembles a panicum; however, *Setaria* species are distinct from panicum in that they almost always have bristles subtending (present just below) the spikelets. Close examination will reveal a few very short bristle awns growing from the base of the seeds. Other common name: bristle panicum.

**Livestock and Wildlife Value:** Reverchon bristlegrass provides good grazing value for livestock and little or no grazing value for white-tailed deer. Seeds are eaten by birds, and turkeys eat them by "stripping" the seedheads.

**Management:** This grass requires no special management.

Seedhead—
Yellow Indiangrass has a distinct yellow-gold seedhead. Seeds have long, bent, twisted awns.

Individual—A healthy clump of yellow Indiangrass can appear conspicuously bluish green.

Community—This healthy community of yellow Indiangrass was found growing on a cattle ranch, in a well-managed riparian area along East Fork Grape Creek. Photos by George Clendenin, USDA-NRCS.

# Yellow Indiangrass *(Sorghastrum nutans)*
## POACEAE—GRASS FAMILY

**Characteristics:** Native; perennial; warm-season.

**Growth Form:** Bunchgrass, but also forms small colonies; height 24 inches (leaves) to 4 to 6 feet (including seedhead).

**Reproduction:** Seeds and short, scaly rhizomes.

**General Description:** Yellow Indiangrass is a large, leafy grass with a very deep root system, native to many soil types in the region but most commonly found in shallow, rocky areas. This is one of the most attractive of all grasses, producing a beautiful golden plume seedhead in early fall. The grass is uncommon in the region, found mostly in protected areas or on roadsides. It is very rarely found in the western half of the Concho Valley region.

**Livestock and Wildlife Value:** Excellent livestock grazing in spring, summer, and fall; dormant grass also provides some winter grazing. This grass provides little or no grazing value for deer. Indiangrass clumps provide good nesting cover for quail and turkeys. Turkeys have been known to "strip" the seeds from the seedheads in fall. Other common name: Indiangrass.

**Management:** This is one of four major tall grasses that formerly made up the majority of the tallgrass prairie, which extended from central Texas northward through the Great Plains. Because of their lack of knowledge about range management, early ranchers overgrazed these grasslands and much of the productivity of the tall grasses was lost. Where Indiangrass still occurs, it will persist with careful and conservative grazing management. Ranches that practice low to moderate stocking rates and rotational grazing will see an increase in this grass. Seeds are commercially available and are often added to range seeding mixes. Establishment of Indiangrass is slower than that of other grasses, often taking three years to make a significant showing.

Community—This tall dropseed community was photographed in October, 8 months after the entire pasture was burned with a prescribed winter burn back in February. It demonstrates how vigorously desirable plants come back after a prescribed burn. It also shows how important it is to rest pastures completely from livestock for a full growing season immediately after a winter burn. This plant was virtually absent or invisible before the fire and actually increased in frequency and abundance after the fire. To be fair, well-timed precipitation also contributed to this success. Photo taken at cattle ranch, Christoval, Texas. Photos by George Clendenin, USDA-NRCS.

# Tall dropseed *(Sporobolus compositus)* (Sy = *S. asper*)
**POACEAE—GRASS FAMILY**

**Characteristics:** Native; perennial; warm-season.

**Growth Form:** Bunchgrass; height 12 inches (leaves) to 30 inches (including seedhead).

**Reproduction:** Seeds.

**General Description:** Tall dropseed is best identified by the large clumps and the very long, thin-tipped leaves. This grass is very closely related to meadow dropseed; the two grasses are merely botanical varieties of the same species: tall dropseed is var. *compositus* and meadow dropseed is var. *drummondii*. As with many species of *Sporobolus,* the spike-like inflorescences of *S. compositus* are at least partly included within a sheath.

**Livestock and Wildlife Value:** This grass provides some grazing in spring but quickly becomes relatively coarse and unpalatable. It normally receives limited grazing use by livestock unless other grasses are not available. This grass provides little or no grazing value for deer. The large clumps provide good nesting cover for quail.

**Management:** On some soils, this grass responds quickly to improvements in grazing management. Where this grass is common, its forage value can be increased considerably by burning, followed by grazing of tender new growth.

**Seedhead**—Many *Sporobolus* plants start with the seedhead partially or fully enclosed in the leaf sheath. As sand dropseed matures, the seedhead expands into an open panicle, but then at final maturity it changes again and may appear as a long, tightly rolled, straw-colored spike. The flag-leaf that projects away from the seedhead is also a field identifier for a young plant.

**Community**—When mature and dry, sand dropseed turns straw yellow and the seedhead arches over distinctively. It appears significantly different when younger, with an open or partially enclosed panicle. Photos by George Clendenin, USDA-NRCS.

# Sand dropseed *(Sporobolus cryptandrus)*
## POACEAE—GRASS FAMILY

**Characteristics:** Native; perennial; warm-season.

**Growth Form:** Thin bunchgrass; height 15 inches (leaves) to 30 inches (including seedhead).

**Reproduction:** Seeds.

**General Description:** Most commonly found on loamy or sandy loam soils but can also be found on deeper clay loam soils. The best identifying characteristic of this grass is the pronounced tuft of white hairs where the leaf attaches to the stem. Seedhead is often half or more enclosed within the sheath of upper leaves.

**Livestock and Wildlife Value:** Provides fair grazing for livestock during the early growing season. This grass provides little or no grazing value for deer. Seeds are very tiny and are eaten by several species of grassland songbirds. Seedheads are "stripped" by turkeys.

**Management:** This grass sometimes increases quickly in response to improvements in grazing management. Seeds are commercially available and are often added as a minor component of range seeding mixes for quick cover.

**Native American Uses:** Native Americans consumed the seeds of sand dropseed (Diggs, Lipscomb, and O'Kennon 1999).

Seedhead—The typical *Tridens* seedhead has spikelets with 4–12 florets that are stacked in *V*-shaped bracts (glumes). Seedheads of white tridens have a purple tinge when young and then turn a white or straw color at maturity. Photos by George Clendenin, USDA-NRCS.

# White tridens *(Tridens albescens)*
## POACEAE—GRASS FAMILY

**Characteristics:** Native; perennial; warm-season.

**Growth Form:** Bunchgrass; height 15 inches (leaves) to 30 inches (including seedhead).

**Reproduction:** Seeds.

**General Description:** This grass grows in seasonally wet areas where extra water accumulates. The seedheads are fat spikes, crowded with many short seed branches. The leaves and sheaths do not have hair. The stems are also hairless, or the lowermost nodes are sparsely bearded. The plant sometimes produces a musky odor.

**Livestock and Wildlife Value:** White tridens provides limited livestock grazing in the early growing season, but it is not highly preferred. Seeds are an important food for turkeys, which "strip" them from the seedheads.

**Management:** No special management is recommended for this grass, which seldom makes up a significant percentage of forage production.

Seedhead—This photo showing slim tridens with purple-tinged spikelets was actually a rarity for me; it came after well-timed winter and spring precipitation. Usually the seedhead is dried out and straw colored.

Community—Slim tridens is mostly apparent in a pasture, with a very nondescript straw-colored, slender, spike-like inflorescence. On some of our "well-worn" Edwards Plateau ecological sites, there is slim tridens, red grama, hairy tridens, threeawn, and not much else, since all of these grasses are at the bottom of the list of livestock preference. Photos by George Clendenin, USDA-NRCS.

# Slim tridens *(Tridens muticus* var. *muticus)*
## POACEAE—GRASS FAMILY

**Characteristics:** Native; perennial; warm-season.

**Growth Form:** Bunchgrass; height 12 inches (leaves) to 24 inches (including seedhead).

**Reproduction:** Seeds.

**General Description:** There are two varieties of *Tridens muticus* that are recognized in Texas: *T. muticus* var. *elongatus* (rough tridens) and *T. muticus* var. *muticus* (slim tridens), with slight botanical differences between the two. Both are recorded as present in the Concho Valley (Amos 1998). Slim tridens is noted as more widespread throughout Texas, especially to the south and west of our region. Rough tridens is more widespread in north-central Texas (Powell 1994; Gould 1975). The plant grows mostly in shallow, rocky soil. Seedheads are thin spikes with sparse seed-bearing branches.

**Livestock and Wildlife Value:** Slim tridens provides limited grazing for all classes of livestock but is not preferred. Seeds are an important food source for turkeys, which "strip" them from the seedheads. Quail and other grassland birds sometimes eat the seeds. Larger clumps of slim tridens, of sufficient size, provide nesting cover for quail.

**Management:** No special management is needed to maintain this grass.

Seedhead—Male and female flowers are on the same plant. Here the male pollen-producing anthers are on top, while the female part—the long, feathery, pollen-receiving stigma—is at the bottom of the seedhead. Photo by Mark Meyer, USDA-NRCS.

Community—Many sites in our area have tremendous stands of this beautiful grass. Many of these locations are along riparian corridors. This photo was taken along a portion of Spring Creek, where the landowner has protected the riparian area by establishing a riparian buffer. Photo by George Clendenin, USDA-NRCS.

# Eastern gamagrass *(Tripsacum dactyloides)*
## POACEAE—GRASS FAMILY

**Characteristics:** Native; perennial; warm-season.

**Growth Form:** Very large bunchgrass; height 3 feet (leaves) to 6 feet (including seedhead).

**Reproduction:** Seeds and rhizomes.

**General Description:** Grows naturally in rich bottomland soils near creeks and rivers. Seedheads are unmistakable, with male pollen-producing segments at the top and female flowers enclosed in hard burs at the base. This grass produces very thick rhizomes, which helps enlarge the size of individual clumps.

**Livestock and Wildlife Value:** This excellent grass may be the very finest forage grass of the region. Eastern gamagrass provides large quantities of very high-quality forage during the growing season and usually retains some green leaves into early winter if temperatures are mild.

**Management:** This grass can be found in bottomlands that have been rested from grazing or that have a long history of very good grazing management. Since this grass is so palatable to livestock, it is easily overgrazed unless the manager takes special measures. The most successful cases of establishment of eastern gamagrass in the Concho Valley have occurred where riparian areas are fenced off and livestock access is managed. Seeds are commercially available and are sometimes planted in pure stands on deep soils and intensively managed for forage production or hay. This grass has a very dense, stabilizing root system that is able to withstand the forces of flooding and erosion in riparian areas.

Flowers are tubular and usually white, although they can be purple on separate plants. Photos by George Clendenin, USDA-NRCS.

FORBS

# Ruellia *(Ruellia metziae)*
ACANTHACEAE — WILD PETUNIA FAMILY

**Characteristics:** Native; perennial.

**Flower Bloom Period:** Spring–fall in response to rainfall.

**Growth Form:** Upright forb with multiple stems; 12–16 inches tall when flowering.

**Reproduction:** Prolific seed producer. Seedpods "explode" at maturity, throwing seeds some distance from the plant.

**General Description:** Flowers are tubular, splitting into 5 petals. Flowers are usually white but can also be purple on separate plants. Ruellia is most commonly found in deep soils; it is infrequent in shallow or rocky soils. The plant has large leaves with irregular or toothed edges, and leaves form in opposite pairs on the stem. Other common name: wild petunia.

**Livestock and Wildlife Value:** Ruellia provides good feed for sheep, goats, and white-tailed deer in spring and is sometimes eaten by cattle. Seedpods are presumably eaten by turkeys.

**Management:** Like many other desirable perennial forbs, this plant is best maintained by conservative grazing management.

These odd-looking flowers are unique in the Concho Valley, with 3 petals pointing downward and 1 pointing upward. This lower 3-lobed "lip" typically has a white spot with purple dots at the base. The lower "tube" of the flower is covered in dense hairs. Photos by George Clendenin, USDA-NRCS.

# Hairy tube tongue *(Siphonoglossa pilosella)*
## ACANTHACEAE — WILD PETUNIA FAMILY

**Characteristics:** Native; perennial.

**Flower Bloom Period:** Spring–fall in response to rainfall.

**Growth Form:** Grows in low, dense clumps; diameter up to 12 inches, height up to 6 inches.

**Reproduction:** Seeds and rhizomes; most plants in this family have seedpods that "explode" when ripe, throwing the seeds some distance from the plant.

**General Description:** Hairy tube tongue can be found most commonly in deeper soils and is frequently found under the dappled shade of mesquite, which provides a favorable environment. Small oval leaves are formed in opposite pairs along the stem and help to identify this plant when no flowers are present.

**Livestock and Wildlife Value:** An extremely desirable plant for sheep, goats, and white-tailed deer, also grazed to a lesser extent by cattle.

**Management:** Because this plant is so palatable, it is scarce in many places. Hairy tube tongue is often found growing in or near the protection of prickly pear, spiny shrubs, or brush piles. It responds favorably to good conservative grazing management and proper control of white-tailed deer numbers.

Photos by George Clendenin, USDA-NRCS.

# Purple eryngo *(Eryngium leavenworthii)*
## APIACEAE — CARROT OR PARSLEY FAMILY

**Characteristics:** Native; annual.

**Flower Bloom Period:** Fall.

**Growth Form:** Upright forb, branching many times from a single-stemmed base; height to about 24 inches depending on moisture.

**Reproduction:** Seeds.

**General Description:** Superficially resembles a thistle because of the abundant sharp, stiff spines, but it is unrelated to true thistles (which are in the Asteraceae family). New leaves and stems are silver colored. At maturity in fall, the seedheads and leaves turn bright purple.

**Livestock and Wildlife Value:** Young tender leaves are occasionally consumed by sheep, goats, and white-tailed deer before they become spiny and unpalatable. The strange-looking white seeds are occasionally eaten by quail and other seed-eating birds. Purple eryngo is sometimes used in dried flower arrangements because of its striking and attractive purple seedheads.

**Management:** Annuals such as eryngo often grow in greatest abundance in areas of bare ground or sparse vegetation. Maintenance of good grass cover will reduce the abundance of annual forbs, while a thin grass cover invites the germination and growth of annuals.

Fruit—Fruit is a round capsule. Seeds are flat and triangular. Photos by George Clendenin, USDA-NRCS.

# Dutchman's pipe *(Aristolochia coryi)*
## ARISTOLOCHIACEAE—PIPEVINE OR BIRTHWORT FAMILY

**Characteristics:** Native; perennial.

**Flower Bloom Period:** Spring–summer.

**Growth Form:** Low-growing, inconspicuous forb with long, weak stems that trail on the ground. Grows from a large central taproot.

**Reproduction:** Seeds.

**General Description:** Distinctively shaped leaves are roughly triangular with large rounded lobes and are alternately arranged along stem. Flowers are extremely unusual, with a long, curved tube flared at the end. Flat seeds are tightly stacked within the round seedpod. Plant is somewhat uncommon and easily overlooked.

**Livestock and Wildlife Value:** Dutchman's pipe is sometimes eaten by sheep and white-tailed deer. This forb is the larval host plant for the pipevine swallowtail butterfly. The caterpillars of this large attractive butterfly feed only on the leaves of this plant.

**Management:** No specific management practices have been identified for this plant. Avoid overgrazing and use herbicides carefully to help maintain this interesting plant.

Flower—The flower clusters are all attached and radiate from the same point, much like an umbrella. The 5 white "balls" on the flower are called hoods. This species of milkweed attracts many different insects as pollinators.

Seed—The seedpods superficially resemble the miniature horns of pronghorn antelope, thus giving rise to the common name "antelope horn." Photos by George Clendenin, USDA-NRCS.

# Antelope horn milkweed *(Asclepias asperula)*
## ASCLEPIADACEAE—MILKWEED FAMILY

**Characteristics:** Native; perennial.

**Flower Bloom Period:** Spring.

**Growth Form:** Robust forb; height 12 inches, with sprawling stems supported by a very large taproot. Clumps may be up to 24 inches across.

**Reproduction:** Light silky fluff attached to the seeds allows them to be spread by the wind.

**General Description:** Flowers are strange and unusual, forming in large clusters. Leaves are long and narrow, with pointed tips and milky sap, and are alternately arranged along the stem, except possibly the lowest ones. Seedpods are large, elongated capsules that turn dark when nearing maturity. When mature, the capsules split open and release the hundreds of seeds with fluffy hairs.

**Livestock and Wildlife Value:** Animals dislike the taste of milkweeds. They are rarely consumed by livestock or white-tailed deer because of their bitter sap. This and other milkweeds are the larval host plants of the monarch butterfly and queen butterfly. Females lay eggs on the leaves and the caterpillars eat milkweed leaves exclusively. These caterpillars are reportedly distasteful if not toxic to potential predators such as birds. Some songbirds reportedly use the silky hairs to line their nest.

**Management:** To avoid toxicity problems for livestock, graze conservatively and always leave a surplus of grass in the pasture.

**Toxic Agent:** Cardiac glycosides. A toxic dose is generally considered to be 1.2 percent of an animal's weight in green plant material (Hart et al. 2010).

**Toxic Description:** Poisons all livestock, especially sheep. Signs appear within a few hours of ingestion of a toxic dose, and death usually follows within a few days in most fatal cases. Antelope horn milkweed is especially prevalent in west central Texas and the Concho Valley. The plant is most toxic before it matures and less toxic as it dries. It can still be dangerous in hay. See *Toxic Plants of Texas* (Hart et al. 2010) for livestock clinical signs and more detailed information.

Fruit—Large, elongated seedpods break open at maturity to release the seeds with hairs. Photos by George Clendenin, USDA-NRCS.

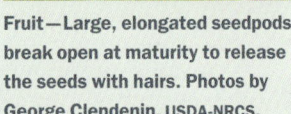

# Broadleaf milkweed (Asclepias latifolia)
## ASCLEPIADACEAE — MILKWEED FAMILY

**Characteristics:** Native; perennial.

**Flower Bloom Period:** Spring.

**Growth Form:** Large, upright, robust forb with a large taproot; height to 18 inches.

**Reproduction:** Light silky fluff attached to the seeds allows them to be spread by the wind.

**General Description:** Leaves are large, round, and thick, with milky sap. Clusters of white flowers are formed along the stalk at the base of the upper leaves.

**Livestock and Wildlife Value:** Animals dislike the taste of milkweeds. They are rarely consumed by livestock or white-tailed deer because of their bitter sap. This and other milkweeds are the important larval host plants of the monarch butterfly and queen butterfly. Female butterflies lay eggs on the leaves and the caterpillars eat milkweed leaves exclusively. These caterpillars are reportedly toxic to potential predators such as birds.

**Management:** To avoid toxicity problems for livestock, graze conservatively, always leaving a surplus of grass in the pasture.

**Toxic Agent:** Cardiac glycosides. A toxic dose for cattle is generally considered to be 1.0 percent of their body weight in broadleaf milkweed, and amounts as low as 0.15 percent have poisoned sheep and goats (Hart et al. 2010).

**Toxic Description:** Poisons cattle and goats, but sheep are the most susceptible. Signs appear within a few hours of ingestion of a toxic dose, and death usually follows within a few days in most fatal cases. Especially prevalent in west central Texas and the Concho Valley. Broadleaf milkweed is toxic in all stages of growth but is most toxic when immature. See *Toxic Plants of Texas* (Hart et al. 2010) for livestock signs and more detailed information.

**Spring Blooms (February–June)**—One of the first blooms in the Concho Valley after a wet winter. Yellow ray flowers number 8–10, with 3–4 notches at each tip.

**One of a Kind**—Huisache daisy is in a monotypic genus (*Amblyolepis*), meaning it's the only member of this genus because of its unique characteristics. It is also endemic (unique and native to) the southwestern United States and Mexico. Notice the long, cottony hairs along the stem and leaves.

**Community**—Huisache daisy grows in large colonies. Photos by George Clendenin, USDA-NRCS.

# Huisache daisy *(Amblyolepis setigera)*
## ASTERACEAE—SUNFLOWER OR DAISY FAMILY

**Characteristics:** Native; annual.

**Flower Bloom Period:** Early spring.

**Growth Form:** Small forb; height 8–16 inches when flowering.

**Reproduction:** Seeds.

**General Description:** Fragrant, sweet-smelling flowers. Huisache daisy grows in large colonies. Stems and leaf margins are covered with long, cottony hairs. Leaves are clasping on the stem.

**Livestock and Wildlife Value:** Huisache daisy is readily grazed by sheep and goats during a brief period in the spring.

**Management:** This and most other annual forbs grow best in areas with sparse grass cover and are abundant only in wet years. An abundance of "winter weeds" provides short-lived forage but also may indicate a lack of adequate grass cover.

This is the most common and widespread species of ragweed in the region and is often confused with western ragweed (*Ambrosia psilostachya*). Field ragweed has much larger and more intricately divided leaves than western ragweed. See chapter 4, "Comparing and Contrasting Similar Plants."

The leaf shape is pinnatifid because it resembles a pinnately compound leaf. Photos by George Clendenin, USDA-NRCS.

### What Makes Some Pollen Allergenic?

Not all types of pollen cause hay fever. Several things need to occur before a plant becomes an allergy problem:

1. The pollen must be loosely attached to the plant, so that it will float a considerable distance in the wind. In general, wind-pollinated plants have more buoyant pollen than insect-pollinated plants. Many insect-pollinated plants do not need to produce abundant pollen to ensure effective pollination. Ragweeds are wind pollinated and have abundant pollen.
2. The pollen must be produced in abundance. Several plants produce a lot of pollen but the pollen is not allergenic. The prolific pollen-producing pines of East Texas may be an example of this. Strike two for ragweed, an abundant pollen producer!
3. The pollen grains must irritate the sinuses. This sounds simple, but a plant may satisfy criteria 1 and 2 above and still not bother you. Some do and some don't, regardless of the buoyancy or abundance alone. Ask your allergy doctor for help in identifying the plants that bother you as an individual.

# Field ragweed *(Ambrosia confertiflora)*
## ASTERACEAE—SUNFLOWER OR DAISY FAMILY

**Characteristics:** Native; perennial.

**Flower Bloom Period:** Fall.

**Growth Form:** Colony-forming forb; colonies may be very large and dense or scattered and sparse. Individual plants are normally 8–16 inches tall depending on rainfall.

**Reproduction:** Seeds and rhizomes.

**General Description:** The male flowers form at the top of the plant on spikes and produce large amounts of pollen, which is carried by the wind. The female flowers are nearly invisible and located on the lower part of the plant where the leaves attach to the stem; this is also where the seeds form. The flowers of the *Ambrosia* species are very small, abundant, and wind pollinated instead of bee pollinated. The airborne pollen is a leading cause of allergic reactions in the fall and is considered the leading cause of hay fever in the United States. The seeds of field ragweed resemble very tiny grass burs.

**Livestock and Wildlife Value:** This plant is sometimes eaten in small amounts by white-tailed deer, sheep, and goats but is not considered a desirable forage plant. The seeds of field ragweed are of little or no use by birds as compared to those of western ragweed, which offer an excellent food source for quail.

**Management:** Field ragweed increases as desirable grasses decrease. Good range management practices and good grass cover will help keep field ragweed from increasing. Field ragweed can be treated with a broadleaf herbicide containing 2,4-D or similarly labeled product.

**Native American Use:** Native Americans used as a medicinal tea.

Leaf—Western ragweed has much smaller and less intricately divided leaves than field ragweed. This is best seen in a side-by-side comparison in chapter 4, "Comparing and Contrasting Similar Plants." Photos by George Clendenin, USDA-NRCS.

# Western ragweed *(Ambrosia psilostachya)*
## ASTERACEAE—SUNFLOWER OR DAISY FAMILY

**Characteristics:** Native; perennial.

**Flower Bloom Period:** Fall.

**Growth Form:** Colony-forming forb. In the Concho Valley, colonies of western ragweed are normally scattered and sparse. Individual plants are normally 10–20 inches tall depending on rainfall.

**Reproduction:** Seeds and rhizomes.

**General Description:** Western ragweed is far less common than field ragweed in the Concho Valley. Western ragweed has much smaller and less intricately divided leaves than field ragweed. See chapter 4, "Comparing and Contrasting Similar Plants," for a side-by-side comparison. The male flowers form at the top of the plant on spikes and produce large amounts of pollen, which is carried by the wind. The female flowers are nearly invisible and are located on the lower part of the plant where the leaves attach to the stem; this is also where the seeds form. The flowers of the *Ambrosia* species are very small, abundant, and wind pollinated instead of bee pollinated. The airborne pollen is a leading cause of allergic reactions in the fall and is considered the leading cause of hay fever in the United States. In the Concho Valley, western ragweed grows primarily in deeper soils, especially near creeks and draws, and is most commonly found in sandy loam soils.

**Livestock and Wildlife Value:** This plant is grazed sparingly by cattle, sheep, goats, and white-tailed deer during early spring but is otherwise not considered a desirable forage plant. The seeds of western ragweed are excellent food for quail.

**Management:** Landowners who are interested in livestock grazing at the expense of quail habitat will not want to encourage western ragweed and may want to control it with herbicides. Those who wish to accommodate both livestock and wildlife will usually want to retain this plant. It can be encouraged to multiply by disking shallow strips in winter. Seeds are readily available through commercial seed sources, but the success of establishment in the region is unknown.

Flower—Lazy daisy has about 20–40 ray flowers (more consistently in the 20s in the Concho Valley), with the flower head measuring about ¾–1½ inches across.

Lazy daisy is sometimes confused with fleabane, but lazy daisy has larger flowers and fewer petals (ray flowers) than fleabane. Photos by George Clendenin, USDA-NRCS.

# Lazy daisy *(Aphanostephus skirrhobasis)*
## ASTERACEAE—SUNFLOWER OR DAISY FAMILY

**Characteristics:** Native; perennial or annual.

**Flower Bloom Period:** Spring.

**Growth Form:** Small, bushy forbs; height 6–12 inches.

**Reproduction:** Seeds.

**General Description:** The genus *Aphanostephus* contains 4 species, all endemic to the United States and Mexico. All 4 species have flowers that droop and remain closed until midmorning or later and then open and remain open at night. Three species of lazy daisy are found in this region. They have a similar appearance and are combined in this description. Two are annuals, while one is perennial. Lazy daisy is covered by soft, fine hairs. The undersides of unopened flowers appear rose or reddish purple. Other common name: Arkansas lazy daisy.

**Livestock and Wildlife Value:** Lazy daisy is grazed by sheep, goats, and white-tailed deer, especially in the early stage of growth. The seeds are too small to be of value to quail.

**Management:** No specific management is needed or recommended to maintain this species on rangeland. As always, the prevention of overgrazing and limited use of herbicides is the best way to maintain good plant diversity.

Photos by George Clendenin, USDA-NRCS.

# Mexican sagewort *(Artemisia ludoviciana)*
## ASTERACEAE — SUNFLOWER OR DAISY FAMILY

**Characteristics:** Native; perennial.

**Flower Bloom Period:** Fall.

**Growth Form:** Forb/herb, forms small to large colonies; height 12–24 inches depending on soil and rainfall, but plants often lean or fall over by late fall.

**Reproduction:** Spreads primarily by rhizomes and root sprouts and less commonly by seeds.

**General Description:** Mexican sagewort is closely related to sagebrush and has the characteristic sage smell. In the fall, the inconspicuous flowers form and the plant releases pollen. Leaves are very variable in terms of shape and pubescence (hairs). Young leaves are white and woolly above and below, to dark and hairy above and white and woolly below, which gives the plant a silvery color. Like *Ambrosia* species, *Artemisia* species are also wind pollinated and cause allergies. However, because they are less abundant in the Concho Valley, they are not a problem allergen. The plant begins new growth in winter and continues to grow all during spring and summer. It remains alive after frost but then begins to turn gray and the new root sprouts for the next year begin to emerge from the ground.

**Livestock and Wildlife Value:** This plant is often found on roadside rights-of-way and less frequently in grazed pastures, indicating use by livestock. While not a highly preferred forage, it is used by sheep, goats, and deer in late winter and early spring when other green forage is in short supply.

**Management:** Like most other desirable forbs, this plant is best maintained by conservative and careful grazing management practices and proper management of white-tailed deer numbers. This plant has some potential value in native landscaping because of its unique and attractive color and its high drought tolerance.

Flower—Fall Blooms (September–October). The numerous crowded heads are up to ½ inch across. There are up to 15 white ray flowers, and yellow disk flowers. *Aster* in Greek means "a star," describing the radiant heads. Photos by George Clendenin, USDA-NRCS.

# Heath aster *(Aster ericoides)*
## ASTERACEAE—SUNFLOWER OR DAISY FAMILY

**Characteristics:** Native; perennial.

**Flower Bloom Period:** Fall.

**Growth Form:** Dense, colony-forming forb; normal height 15–30 inches depending on rainfall.

**Reproduction:** Spreads primarily by rhizomes and less commonly by seeds.

**General Description:** The abundant flowers make this native plant very attractive and desirable. It is usually found along roadsides and fence lines in the Concho Valley. This evergreen plant begins growth in late winter as new root sprouts. In the early stages of growth, the plant is very leafy. By midspring, it begins to grow upright, with many thin stems and small leaves. By summer the plant has become more stemmy and less leafy. In the fall, hundreds or thousands of small white daisy flowers form. After flowering, the plant forms the feathery seeds in late fall. By winter the old growth becomes ragged looking but new plants begin to form again from the root sprouts.

**Livestock and Wildlife Value:** Heath aster is a highly preferred Class I plant for white-tailed deer. See the section titled "Managing for White-Tailed Deer in West Central Texas" in chapter 6 for a description of the classes of browse preference.

**Management:** This plant is part of the natural diversity of the ecological community and does not pose a threat.

(left) Flowers—Bloom in spring and persist into summer. Heads 1–2 inches across. Yellow ray flowers (usually 8), yellow disk flowers.

(right) Leaves—Leaves are mostly basal and deeply cut or lobed. Upper leaves are smaller and also lobed.

Sneak-a-Peek—This basal rosette of leaves appears in winter and early spring, revealing the presence of this desirable plant.

Changing Flowers—The yellow ray flowers are known to curl under or downward during intense heat or sunlight during the day and expand back afterward. Photos by George Clendenin, USDA-NRCS.

# Engelmann daisy *(Engelmannia peristenia)*
## ASTERACEAE — SUNFLOWER OR DAISY FAMILY

**Characteristics:** Native; perennial.

**Flower Bloom Period:** Early spring–early summer.

**Growth Form:** Robust bushy forb with large taproot; height 12–24 inches depending on soil and rainfall.

**Reproduction:** Seeds.

**General Description:** Engelmann daisy is easily recognized by its long, deeply cut or lobed leaves and tall, erect flowering stems. Leaves are covered by short, stiff hairs that may become rough and bristly by summer. In the winter, a new cluster of leaves emerges from the taproot. By early spring, the leaf cluster begins to grow upright into flowering stalks. Flowers begin to form in early spring and may continue to bloom into early summer depending on rainfall. In some wet years, the plant blooms all summer long.

**Livestock and Wildlife Value:** Engelmann daisy is highly preferred by livestock and white-tailed deer, leading ranchers and range scientists to sometimes refer to it as an "ice cream" plant.

**Management:** Since this plant is highly preferred by livestock, it is seldom seen in grazed pastures. Proper range management will ensure that it remains part of the native mix. Seeds are also commercially available through native seed suppliers.

Flower—Although similar in appearance to lazy daisy, fleabane has more abundant and crowded ray flowers, over 40 typically. There are also differences in the yellow disk flowers—those of lazy daisy are smaller. Photos by George Clendenin, USDA-NRCS.

# Fleabane *(Erigeron strigosus)*
## ASTERACEAE—SUNFLOWER OR DAISY FAMILY

**Characteristics:** Native; commonly biennial, sometimes annual or perennial.

**Flower Bloom Period:** Spring.

**Growth Form:** Small, weak-stemmed forb from a taproot; central stalk branches many times to form a "bushy" plant that grows to a height of about 12 inches.

**Reproduction:** Seeds.

**General Description:** Fleabane, a nondescript forb, is easily overlooked when no flowers are present. When the flowers appear in spring, it can be readily identified by the dozens of small daisy-like flowers. Tiny hairs cover this plant, and it has many linear leaves. Other common name: prairie fleabane.

**Livestock and Wildlife Value:** Consumed by sheep, goats, and white-tailed deer.

**Management:** Sometimes found growing near the protection of prickly pear, indicating that it may be somewhat susceptible to heavy grazing. Prickly pear and spiny shrubs are often underappreciated for their role in protecting these and other plants.

Rabbit tobacco shows a cluster of white heads in the spring, with green leaves surrounding them.

It is more common to see rabbit tobacco like this, during the summer and fall, after it has dried up and died. It may persist in this state for months. Photos by George Clendenin, USDA-NRCS.

# Rabbit tobacco *(Evax verna)*
## ASTERACEAE—SUNFLOWER OR DAISY FAMILY

**Characteristics:** Native; annual.

**Flower Bloom Period:** Spring.

**Growth Form:** Very small taprooted forb, only a few inches tall.

**Reproduction:** Seeds.

**General Description:** Current year's plants are silvery gray in appearance because of a thick mat of hairs on the leaves and stem. Rabbit tobacco usually has a single stem ending with a white cluster of heads, or it branches into two stems, with a white cluster of heads at the end of each stem and in the axils. Old dead plants may persist but turn dark gray. Flowers are completely inconspicuous.

**Livestock and Wildlife Value:** There is no documented value to livestock or white-tailed deer. Songbirds sometimes gather the plants for nest material.

**Management:** This odd plant is extremely common in some years and totally absent in other years, depending on rainfall. All native plants have some value to the ecosystem, even if not consumed by livestock or wildlife, but we have not yet discovered the hidden virtues of some plants, such as this one.

**An Odd-Looking Flower Head**—Pincushion daisy may be seen with or without the yellow ray flowers. When present, the ray flowers are 3-lobed and purple veined on their underside. There may be 6–13 ray flowers. The disk flowers form a ¾-inch sphere in the center of the flower head and are deep purplish red to brown.

Pincushion daisy ranges from northern Mexico into the Trans-Pecos region of Texas, and into Oklahoma, Colorado, and Utah. Photos by George Clendenin, USDA-NRCS.

# Pincushion daisy *(Gaillardia pinnatifida)*
## ASTERACEAE — SUNFLOWER OR DAISY FAMILY

**Characteristics:** Native; perennial.

**Flower Bloom Period:** Spring.

**Growth Form:** Before the flowers form, the plant is a small inconspicuous forb, or as some say, "just another weed." Flowering height is 12–18 inches.

**Reproduction:** Seeds.

**General Description:** The cluster of leaves at the base can be extremely variable. Sometimes the leaves are broad and rounded, sometimes long and narrow, and sometimes highly dissected with many intricate lobes. Upon flowering, the plant sends up one to several tall stalks, each with a single flower head. Flowers may or may not have petals. Most flower heads consist of a "ball," which is actually a cluster of many tiny flowers. The balls may be reddish or silvery depending on age. They are strongly and sweetly fragrant. Other common names: yellow gaillardia, slender gaillardia.

**Livestock and Wildlife Value:** Leaves and flower heads are consumed by sheep, goats, and white-tailed deer, and occasionally cattle.

**Management:** Proper management of deer and livestock numbers will allow for the perpetuation of this and other desirable forbs.

Back of flower. Photos by George Clendenin, USDA-NRCS.

Flower—Unlike any other wildflower in its color. If you look closely at the yellow-tipped petals (ray flowers), you will notice that they are actually 3-lobed and are not just individual ray flowers.

# Indian blanket *(Gaillardia pulchella)*
## ASTERACEAE—SUNFLOWER OR DAISY FAMILY

**Characteristics:** Native; annual.

**Flower Bloom Period:** Spring–early summer.

**Growth Form:** A weedy-looking plant prior to blooming; after blooming, a striking and attractive wildflower.

**Reproduction:** Seeds.

**General Description:** One of the more popular and easily recognized wildflowers in the Concho Valley. The characteristic red and yellow petals (ray flowers) are not found on any other wildflower in the region. Other common names: firewheel, rose-ring gaillardia.

**Livestock and Wildlife Value:** Plants are eaten in early spring by sheep, goats, and white-tailed deer and are sometimes grazed by cattle.

**Management:** Annual forbs are usually more abundant in areas with sparse grass. Heavy grazing or other disturbance often promotes the growth of many annual species. A good dense cover of grass reduces the opportunity for these plants to grow. An abundance of wildflowers often, but not always, indicates a lack of good grass cover. Those who wish to develop wildflower areas can purchase seeds of this and many other species of native flowers. Wildflower seeds are typically planted in the fall.

Sticky Situation—Sometimes confused with sawleaf daisy, curlycup gumweed has much smaller leaves with less prominent serrations, and markedly different seeds. Photos by George Clendenin, USDA-NRCS.

# Curlycup gumweed *(Grindelia nuda)*
## ASTERACEAE—SUNFLOWER OR DAISY FAMILY

**Characteristics:** Native; annual or biennial.

**Flower Bloom Period:** Spring–summer.

**Growth Form:** Multistemmed forb from a taproot; height 18–24 inches.

**Reproduction:** Seeds.

**General Description:** Leaves and flower heads are covered by a gummy resin that sometimes gives the plant a sticky texture. The flower heads may or may not have petals (ray flowers). The small leafy bracts under the flower head sometimes curl downward.

**Livestock and Wildlife Value:** Generally not consumed by livestock or white-tailed deer except in extreme situations. Any grazing on this plant indicates overgrazing or excessive white-tailed deer numbers. Seeds are eaten by quail and some songbirds.

**Management:** This and other perennial forbs of low value are best kept under control by conservative grazing management that retains a good cover of grass.

**Native American Uses:** "Among the Teton Dakota a decoction of the plant was given to children as a remedy for colic. A Ponca said this was given also for consumption. The tops and leaves were boiled, according to a Pawnee informant, to make a wash for saddle galls and sores on horses' backs" (Gilmore 1919).

Photos by George Clendenin, USDA-NRCS.

# Sawleaf daisy *(Grindelia papposa)*
## ASTERACEAE — SUNFLOWER OR DAISY FAMILY

**Characteristics:** Native; annual.

**Flower Bloom Period:** Late summer–fall.

**Growth Form:** Tall, stout forb; height 2–4 feet depending on soil conditions and rainfall.

**Reproduction:** Seeds.

**General Description:** As suggested by the plant's name, the leaves have a conspicuous saw-toothed edge. The base of the leaves clasps the stem, providing a good clue for identification.

**Livestock and Wildlife Value:** Livestock and white-tailed deer rarely, if ever, graze this plant. The seeds are frequently eaten by quail and other grassland birds.

**Management:** Found more commonly in disturbed areas, this plant is not common in well-managed ranges with good grass cover.

**Plant ID Clue**—Broomweed is sometimes confused with broom snakeweed in the field. One way to distinguish them is to look at the shape of the plant. Broomweed, an annual, branches and blooms ²⁄₃ of the way up the stalk. Thus you have the broom "handle" being the stem, and the "broom" being the branched seedhead. Broom snakeweed, a perennial, branches and blooms from the base.

**A Broomweed Year**—Ranchers know that their pastures can go several years without a significant presence of broomweed, and then all of the sudden, bam, they have a "broomweed year." This pasture looks like it is having a broomweed year! Photos by George Clendenin, USDA-NRCS.

# Common broomweed *(Gutierrezia dracunculoides)*
## ASTERACEAE—SUNFLOWER OR DAISY FAMILY

**Characteristics:** Native; annual.

**Flower Bloom Period:** Late summer–early fall.

**Growth Form:** When mature, a single-stalked weed that branches many times to form a large, dense "broom head" top; height 18–24 inches.

**Reproduction:** Seeds.

**General Description:** Plant begins to bolt upward in early spring as a single slender green stem. By late spring, the stem begins to branch and rebranch, starting about 6 or 8 inches above the ground. By late summer, an intricately branched canopy of green stems is formed and hundreds of small yellow flowers appear. At first frost, the plant dies, but the semiwoody stems persist through the winter. Other common name: annual broomweed.

**Livestock and Wildlife Value:** Broomweed is often underrated. The tips of young plants are readily eaten by sheep and white-tailed deer in spring. The seeds are commonly eaten in large quantity by quail and grassland birds. After the plants have branched to form the broom, they provide surprisingly good overhead cover for quail. Quail can walk about under the canopy of broomweed searching for food and remain hidden from hawks. Despite its considerable value, most ranchers consider broomweed a nuisance weed. A heavy cover of broomweed competes with grass growth and reduces forage production. After the plants die, the persistent stems of dead broomweed form a dense network of rigid sticks that have been known to damage the eyes of livestock.

**Management:** The best way to manage broomweed populations is to keep a good dense cover of grass, which will reduce infestations. Many ranchers will strategically graze broomweed areas with sheep in the spring, reducing the plant population and problems later in the year. Prescribed burning in late winter or early spring can also be used to reduce broomweed infestations.

A Relaxing Bath—Escoba de la vibora (broom snakeweed and other *Gutierrezia* species) is used by the Spanish-speaking peoples of New Mexico and Arizona as a pleasant, therapeutic hot bath for the treatment of arthritis, rheumatism, sore muscles, and hyperextensions. Clip the flowering stems and tie them together in small bundles. Steep a bundle in a quart of boiling water for one-half hour, cool, remove the bundle (discard), and add the water to a hot bath. Enjoy as often as desired! (Moore 1989). Photos by George Clendenin, USDA-NRCS.

# Broom snakeweed *(Gutierrezia sarothrae)*
## ASTERACEAE—SUNFLOWER OR DAISY FAMILY

**Characteristics:** Native; perennial.

**Flower Bloom Period:** Fall.

**Growth Form:** Often described as a subshrub, this plant resembles a miniature bush. Grows from a central woody stalk and taproot, branching to form a dense broom top to a height of 12–18 inches.

**Reproduction:** Seeds.

**General Description:** This plant is much stouter and usually shorter than its cousin, common broomweed. The small leaves grow all winter long, crowded along the woody stems. Flower stalks begin to elongate in summer and bloom in late summer or early fall with thousands of tiny yellow flowers. The leaves and young stems are sometimes sticky with resin. Leaves have a distinctive smell when crushed, giving the plant the name "turpentine weed." It grows more commonly in shallow soil where grass cover is sparse. Other common name: perennial broomweed.

**Livestock and Wildlife Value:** Broom snakeweed can be toxic to livestock; see below. White-tailed deer readily consume this plant in winter when other green plants are scarce, and it causes no known health problems. Wild animals can often detoxify poisonous plants in the digestion process. The seeds are eaten by quail and other grassland birds.

**Management:** Avoid grazing cattle on sandy soils in broomweed-infested pastures; see information on toxicity below. This is a short-lived cyclic plant, being very prolific for several years and then dying back, often because of insect damage to the taproot. It can be kept under control by a good grass cover and allowed to increase by heavy grazing. Prescribed burning can provide good control of broom snakeweed. It can also be controlled with herbicides, but the expense is difficult to justify.

**Native American Uses:** "A decoction of the herb was given to horses as a remedy for too lax a condition of the bowels. They were

induced to drink the bitter preparation by preventing them access to any other drink" (Gilmore 1919).

**Toxic Agent:** Not known, but may be steroidal saponin. However, the toxicity is noted on sandy soils, not the clay loam soils common to the Concho Valley. The plant may accumulate selenium when on high-selenium soil (Hart et al. 2010).

**Toxic Description:** Broom snakeweed poisons cattle, sheep, goats, and swine. It is most toxic at earlier growth stages, usually in late winter or early spring. Cattle abort after eating as little as 20 pounds of fresh broomweed in 7 days. Cattle, sheep, and goats have been killed by eating 10–20 percent of their body weight over a span of 2 weeks. See *Toxic Plants of Texas* (Hart et al. 2010) for livestock clinical signs and more detailed information.

Photos by George Clendenin, USDA-NRCS.

# Sticky selloa *(Gymnosperma glutinosum)*
## ASTERACEAE—SUNFLOWER OR DAISY FAMILY

**Characteristics:** Native; perennial.

**Flower Bloom Period:** Fall.

**Growth Form:** Woody at the base; height 18–30 inches depending on soil and rainfall.

**Reproduction:** Seeds.

**General Description:** Sticky selloa is, well, sticky. It shares its sticky properties with curlycup gumweed. It is similar in appearance to broom snakeweed, but much taller and less compact. Stems can become reddish as they mature. Leaves are alternate and linear, with smooth margins. Leaves have 3 prominent veins on the lower side, assisting in identification. Found on limestone ledges, steep hillsides, and rocky outcrops of the Edwards Plateau, and in arid grasslands of the Trans-Pecos and Rio Grande Plains. Other common names: gumhead, gumweed, tatalencho.

**Livestock and Wildlife Value:** Browsed to some extent by sheep, goats, and white-tailed deer, especially during dry years when other, more desirable plants are in short supply.

**Management:** This plant is short-lived, cyclic in abundance, and not influenced much by management.

Flower—Gray golden aster has yellow ray flowers and yellow disk flowers and blooms pretty much all summer. Photos by George Clendenin, USDA-NRCS.

# Gray golden aster *(Heterotheca canescens)*
## ASTERACEAE—SUNFLOWER OR DAISY FAMILY

**Characteristics:** Native; perennial.

**Flower Bloom Period:** Summer.

**Growth Form:** Low, sprawling forb from a semiwoody base, forming small loose colonies; height 4–8 inches. Most often found in very shallow or rocky soils.

**Reproduction:** Seeds and rhizomes.

**General Description:** Leaves are alternate, linear shaped, covered by hairs, and a distinctive gray-green color. Crushed leaves often have the faint aroma of camphor.

**Livestock and Wildlife Value:** The chemical composition of the leaves renders this plant unpalatable to livestock and white-tailed deer.

**Management:** This and other nonpalatable forbs are part of the natural mix of native plants. Good grazing management and good grass cover will help minimize the spread of these plants that have little direct benefit to ranchers.

Leaves of Chalk Hill woolly-white are extremely variable, starting with a basal rosette in winter with lobed or cut leaves, then growing smaller as they go up the stem, yet they are still lobed.

Flowers—Ray flowers are lacking, yet there are prominent disk flowers covered by white bracts. Apparently it is rare for the flowers to be solid white, as they are typically red tinged. Based on the photos, our Concho Valley variety appears to be atypical in color.

Some flowers on the same plant can appear different, although they are not. This photo shows some flowers that are still unopened and therefore do not have the disk flowers showing, while others have opened and have the spike-like disk flowers in full view. Photos by George Clendenin, USDA-NRCS.

# Chalk Hill woolly-white
*(Hymenopappus artemisiifolius)*
### ASTERACEAE — SUNFLOWER OR DAISY FAMILY

**Characteristics:** Native; biennial.

**Flower Bloom Period:** Spring.

**Growth Form:** During winter, a low cluster of leaves; in spring, a tall, slender, upright forb; height 18–24 inches.

**Reproduction:** Seeds.

**General Description:** Winter rosette of lobed leaves is covered by dense fuzz, giving the leaves a silver color. As the flower stalk begins to elongate, new leaves lack the dense fuzz and silver color and are much smaller. Flowers are unusual and not particularly attractive. Other common names: ragweed woolly-white, wild cauliflower.

**Livestock and Wildlife Value:** Plant is eaten by sheep, goats, white-tailed deer, and sometimes cattle.

**Management:** The abundance of this and many other forbs on roadsides and their scarcity in adjacent grazed pastures indicates their use by livestock. The study of plants inside and outside grazed pastures provides good information on the relative value of many plants. Conservative grazing management is usually the best way to ensure good plant diversity, as well as healthy range and habitat conditions.

Photos by George Clendenin, USDA-NRCS.

# Gayfeather *(Liatris punctata)*
## ASTERACEAE—SUNFLOWER OR DAISY FAMILY

**Characteristics:** Native; perennial.

**Flower Bloom Period:** Fall.

**Growth Form:** Stiff, upright, multistemmed forb from a large tuber; height 12–24 inches.

**Reproduction:** Seeds.

**General Description:** This plant does not like deep rich soil; it grows naturally on shallow and rocky sites. It is extremely drought tolerant and does not need much rain to thrive. From spring to late summer, it consists of tall unbranched stalks covered by long, narrow, pointed leaves. In late summer or early fall the purple flowers begin to pop out of the elongated stalk. Other common name: blazing star.

**Livestock and Wildlife Value:** New stems in spring are browsed by sheep, goats, and white-tailed deer. After the stems and leaves mature, the plant becomes coarse and much less palatable. Seeds are eaten by quail and many species of songbird. This attractive fall-blooming forb is especially valuable to butterflies that consume energy-rich nectar from the flowers.

**Management:** Conservative grazing management helps ensure the perpetuation of this plant. Those who wish to use this plant in native landscaping or wildflower areas can purchase seeds or transplant the tubers.

**Cultural Uses:** "In northern Mexico and New Mexico among both Hispanics and Indians, the root is carried or displayed as an amulet to ward off witches' spells and mal de ojo (evil eye)" (Moore 2003). Cachana (gayfeather) root is also used for other medicinal purposes.

Flowers—Flowers are pale lavender to purple, with 8–12 ray flowers. Look closely and you can see 3–5 small notches at the end of each square-tipped ray flower. Flowers are said to open for only a short time in the morning. Photo by Dee Ann Littlefield, USDA-NRCS.

Photo by George Clendenin, USDA-NRCS.

# Texas skeleton plant *(Lygodesmia texana)*
## ASTERACEAE—SUNFLOWER OR DAISY FAMILY

**Characteristics:** Native; perennial.

**Flower Bloom Period:** Spring.

**Growth Form:** Much-branched forb from taproot; height 12–18 inches.

**Reproduction:** Feathery seeds, dispersed by wind.

**General Description:** An unusual and distinctive plant, bearing few if any leaves on the naked green stems. Stems have milky sap when broken.

**Livestock and Wildlife Value:** Plants are readily grazed by sheep, goats, and white-tailed deer; less often by cattle. Butterflies are very fond of the flowers.

**Management:** This desirable forb is best managed and maintained by good grazing management practices.

Flowers—20–30 ray flowers; both ray and disk flowers are yellow. Photos by George Clendenin, USDA-NRCS.

# Cutleaf daisy *(Machaeranthera pinnatifida)*
*(Sy = Haplopappus spinulosus)*
## ASTERACEAE—SUNFLOWER OR DAISY FAMILY

**Characteristics:** Native; perennial.

**Flower Bloom Period:** Spring–summer in response to rainfall.

**Growth Form:** Low-growing forb from a semiwoody base and woody taproot; height only 6–8 inches.

**Reproduction:** Seeds.

**General Description:** This forb is more common in the western part of the Concho Valley and becomes even more common in the Trans-Pecos. Small daisy flowers are about 1 inch in diameter. Bristly hairs are attached to the seed and aid in wind dispersal. The leaves are alternate and uniquely lobed or cut, with almost threadlike tips at the end of the lobes. Other common names: yellow spiny daisy, cutleaf ironplant, spiny ironplant, yerba de la quintana.

**Livestock and Wildlife Value:** Readily grazed by sheep, goats, and white-tailed deer.

**Management:** This and other desirable forbs are subject to competitive demand among sheep, goats, and white-tailed deer. In these situations, management of animal numbers is the best way to ensure the good health and vigor of these plants.

Hardy and Beautiful—Rock daisy is one of our more attractive drought-resistant flowering plants. It continues to flower in the heat of the summer, while other plants are suffering. Its bouquet shape should also encourage its use as an attractive garden planting. Photos by George Clendenin, USDA-NRCS.

# Rock daisy *(Melampodium leucanthum)*
## ASTERACEAE — SUNFLOWER OR DAISY FAMILY

**Characteristics:** Native; perennial.

**Flower Bloom Period:** Spring–fall in response to rainfall.

**Growth Form:** Short, compact forb from a semiwoody base and taproot; height 6–8 inches, clumps up to 12 inches in diameter.

**Reproduction:** Seeds.

**General Description:** Flowers are 1 inch in diameter and are best identified by the white petals (ray flowers) with blunt tips. Leaves are opposite and linear to narrowly oblong, and they appear to curl slightly under at the margins. Plant grows most commonly in shallow soils. Other common names: plains blackfoot, blackfoot daisy.

**Livestock and Wildlife Value:** Readily grazed by sheep, goats, and white-tailed deer; sometimes eaten by cattle.

**Management:** Conservative grazing management and control of white-tailed deer numbers will help perpetuate this plant. This is a popular flower for native landscapes and is commercially available as containerized plants.

Leaf—The leaves of lyreleaf are typically pinnately lobed, with the terminal (end) lobe being the largest. Photos by George Clendenin, USDA-NRCS.

# Lyreleaf parthenium *(Parthenium confertum)*
## ASTERACEAE—SUNFLOWER OR DAISY FAMILY

**Characteristics:** Native; perennial.

**Flower Bloom Period:** Summer–fall.

**Growth Form:** Upright forb; flower stalk height 12–16 inches when fully grown.

**Reproduction:** Seeds.

**General Description:** Lyreleaf parthenium has many relatives in the same genus that bear a close resemblance at first glance. False ragweed (*P. hysterophorus*) is a taller, weedy annual with many branches and deeply cut leaves. Lyreleaf is a short perennial and actually starts with a basal rosette of leaves. Lyreleaf is common in our region and also grows throughout the Trans-Pecos. Other common names: feverfew, Gray's feverfew.

**Livestock and Wildlife Value:** Rarely consumed by livestock or white-tailed deer.

**Management:** This plant sometimes dominates pastures that have been heavily grazed for many years. Plants such as this can be used as indicators of management and can signify the need to reduce animal numbers or to suspend grazing for several years to hasten recovery.

**Flower**—The 3–7 broadly elliptical ray flowers droop gently down but hold on tight around the cylindrical center. They can be reddish brown or yellow, or a two-toned combination of these colors. Photos by George Clendenin, USDA-NRCS.

# Upright prairie coneflower
*(Ratibida columnifera)*
## ASTERACEAE—SUNFLOWER OR DAISY FAMILY

**Characteristics:** Native; perennial.

**Flower Bloom Period:** Spring–early summer.

**Growth Form:** Tall, slender plants from a large woody taproot; height 18–24 inches.

**Reproduction:** Seeds.

**General Description:** A common roadside wildflower and one of the more recognizable wildflowers in the region. The central cone (where the seeds are produced) is about 1 inch tall. The leaves are deeply cut and appear almost threadlike. Other common name: Mexican hat.

**Livestock and Wildlife Value:** Rarely eaten by livestock or white-tailed deer. Seeds have some value to grassland birds.

**Management:** This unpalatable plant sometimes dominates as a weed of disturbed areas. Even though this plant does not have high value to livestock or wildlife, it is a popular wildflower and is sometimes planted for its attractive flowers.

**Native American Uses:** The Zuni word for this plant is *Ya´konakïa* (*ya´ko*, "bile," or any nauseating substance in the stomach; *na´kïa*, "to vomit"). The entire plant was soaked or steeped in water and drunk as an emetic (something that makes you throw up) (Stevenson 1908).

Flower—Each flower is about ¾–1 inch across, with yellow ray and disk flowers. Ray flowers usually total between 8 and 13. Photos by George Clendenin, USDA-NRCS.

# Threadleaf groundsel *(Senecio douglasii)*
## ASTERACEAE — SUNFLOWER OR DAISY FAMILY

**Characteristics:** Native; perennial.

**Flower Bloom Period:** Spring–fall in response to rainfall.

**Growth Form:** Evergreen forb with many stems rising from a woody base and taproot; height to 18 inches.

**Reproduction:** Seeds.

**General Description:** Threadleaf groundsel is a hardy, drought-tolerant, perennial wildflower that blooms throughout the summer in the Concho Valley. Leaves are alternate, numerous, and pinnately divided into 3–7 long, linear, almost threadlike lobes. Stems and leaves are covered with a layer of very fine fuzz, giving leaves a silvery-green tint. *Senecio* can be found throughout the drier areas of the western plains, including western Texas, northern New Mexico, southwestern Oklahoma, and northern Mexico. It can also be found in Arizona, Wyoming, and Nebraska.

**Livestock and Wildlife Value:** This plant is toxic to cattle and horses and to a lesser extent to sheep and goats (which require up to 10 times the amount for the same effect as in cattle and horses). Plant is often grazed by pronghorn antelope with no ill effects. The liver of pronghorn is able to detoxify the alkaloids. See toxicity information.

**Management:** Some ranchers routinely pull or hand grub groundsel as they notice it in pastures. Spot treatment with herbicide can also be used effectively. Losses to toxic plants can almost always be reduced by practicing proper grazing management and keeping a good surplus of grass.

**Toxic Agent:** Pyrrolizidine alkaloids (Hart et al. 2010).

**Toxic Description:** Stress from lack of water causes the plant to increase its alkaloid content. Generally for acute poisoning, cattle and horses must eat a dose of 1 to 5 percent of their body weight in threadleaf groundsel over a few days. This type of poisoning is rare under range conditions. Most losses are from chronic poisonings in which cattle and horses consume as little as 0.25 percent of their body weight. Oddly, often up to 6 months may elapse between consumption of this plant and the appearance of chronic signs. See *Toxic Plants of Texas* (Hart et al. 2010) for livestock clinical signs and more detailed information.

Flower—Ray petals number between 15 and 30 and are yellow to yellow orange. In this picture the disk flowers appear darker yellow orange, but they may be yellow at times as well.

*Simsia* leaves are opposite and triangular, with lobed or serrated margins. The entire leaf is covered by coarse hairs, giving it a sandpapery feel. The leaves are sometimes perfoliate, meaning that the leaf bases of opposite leaves appear to have grown together. Compare these leaves to those of *Wedelia* in chapter 4, "Comparing and Contrasting Similar Plants." Photos by George Clendenin, USDA-NRCS.

# Bush sunflower *(Simsia calva)*
## ASTERACEAE—SUNFLOWER OR DAISY FAMILY

**Characteristics:** Native; perennial.

**Flower Bloom Period:** Spring–fall in response to rainfall.

**Growth Form:** Scraggly, bushy, much-branched forb from an enormous taproot; height to 24 inches. This plant tends to grow up into other shrubs for support and protection.

**Reproduction:** Seeds.

**General Description:** Leaves and stems are covered by bristly hairs, giving them a scratchy texture. This plant is extremely hardy because of its large, deep taproot. It grows mostly on shallow, rocky soil. Bush sunflower will flower and make seeds 2 or 3 times per year based on rainfall. In the Concho Valley, even the trained eye can often confuse bush sunflower with orange zexmenia (*Wedelia texana*). See chapter 4, "Comparing and Contrasting Similar Plants," for handy tips on identifying the two species.

**Livestock and Wildlife Value:** Readily grazed by sheep, goats, white-tailed deer, and to a lesser extent, cattle. *Simsia* is one of the more preferred forbs in the region. Seeds are eaten by some kinds of songbirds and flowers are used by butterflies.

**Management:** This forb responds favorably to good grazing management. Plants that have been confined to protected locations will begin to spread into open areas in response to good management. If this forb is found in open, accessible locations, it is a sign of good management. If it is found growing only in the protection of brush piles, prickly pear clumps, or spiny shrubs it is a sign that livestock or white-tailed deer numbers are excessive. Seeds are commercially available for inclusion in range seeding mixtures, and establishment is usually successful.

**Flower Heads**—These are usually triangular and have golden bristle-like flowers.

**Leaves**—These are 3-nerved (see the 3 lines on the leaf blade?). They are also lanceolate, have small hairs on both sides, and are arranged alternately along the stem. Photos by George Clendenin, USDA-NRCS.

# Tall goldenrod *(Solidago canadensis)*
## ASTERACEAE — SUNFLOWER OR DAISY FAMILY

**Characteristics:** Native; perennial.

**Flower Bloom Period:** Late summer–fall.

**Growth Form:** Tall forb, forms colonies by rhizomes; height 2–4 feet in flowering stage.

**Reproduction:** Seeds and rhizomes.

**General Description:** One of the common plants that grow in creek and river bottoms (riparian areas). This plant is very noticeable in fall when blooming, but unremarkable during spring and summer. Other common names: common goldenrod, Canada goldenrod.

**Livestock and Wildlife Value:** Grazed by sheep, goats, cattle, and white-tailed deer.

**Management:** Riparian areas are some of the most productive and ecologically important areas in the region. A good protective cover of dense vegetation is important along creek banks and floodplains. This is one of the plants that helps protect these areas from erosion during flood events. Preferential management of riparian areas will help maintain or restore desirable vegetation.

**Native American Uses:**

**Omaha Tribe:** "Goldenrod served the Omaha as a mark or sign in their floral calendar. They said that its time of blooming was synchronous with the ripening of the corn; so when they were on the summer buffalo hunt on the Platte River or the Republican River, far from their homes and fields, the sight of the goldenrod as it began to bloom caused them to say 'Now our corn is beginning to ripen at home'" (Gilmore 1919).

**Zuni Tribe:** "The crushed blossoms are put into water and the infusion is drunk to relieve pains through the body; they are also chewed for sore throat, and are considered excellent for both troubles. The medicine belongs to all the people" (Stevenson 1908).

Flower—The ray flowers are golden, usually number 8, and have 3-lobed tips. The disk flowers can be yellow or reddish brown. The whole flower measures about 1½ inches across. Photo by George Clendenin, USDA-NRCS.

# Greenthread *(Thelesperma filifolium)*
## ASTERACEAE—SUNFLOWER OR DAISY FAMILY

**Characteristics:** Native; annual or short-lived perennial.

**Flower Bloom Period:** Spring–early summer.

**Growth Form:** Tall, slender stems; height 18 inches.

**Reproduction:** Seeds and short, underground rooting stems.

**General Description:** Leaves are opposite and divided 1–3 times into very narrow or threadlike segments; thus the common name. Threadlike leaves on the lower portion of the plant can be up to 4 inches long. *Thelesperma* can be confused with *Coreopsis* species in the Concho Valley because both have similar flowers and pinnately divided, linear leaves. In general, *Thelesperma* has a "tighter," partially connected bract (below the ray flowers) and linear disk corolla lobes, while *Coreopsis* has obvious spreading bracts (especially obvious when ray flowers are absent) and triangular to ovate disk corolla lobes.

**Livestock and Wildlife Value:** Grazed by sheep, goats, and white-tailed deer.

**Management:** Conservative grazing management helps maintain this forb.

Photos by George Clendenin, USDA-NRCS.

# Common dogweed *(Thymophylla pentachaeta)*
## ASTERACEAE—SUNFLOWER OR DAISY FAMILY

**Characteristics:** Native; perennial.

**Flower Bloom Period:** Spring–fall in response to rainfall.

**Growth Form:** Short, compact forb from a slightly woody base, forms low-mounded tufts; height usually only 3–6 inches when flowering.

**Reproduction:** Seeds.

**General Description:** Dogweed is pretty common in almost all of our limestone or calcareous sites. It is usually found on our rangelands, even when other plants are absent because of drought or overgrazing. Leaves have a strong and distinctive odor when crushed. Flowers are attractive and small, only ½ inch in diameter. Ray flowers number 8–13 and are yellow. Disk flowers are a dull yellow to light yellow orange. Leaves are opposite, pinnately divided into 3–11 lobes, and stiff to the touch.

**Livestock and Wildlife Value:** Common dogweed is rarely, if ever, eaten by livestock or white-tailed deer. Chemicals in the leaves render the plant very unpalatable.

**Management:** Dogweed grows mostly in barren areas and shallow, depleted soil. Plants that can thrive in such harsh areas help add much-needed organic matter to the soil.

Flower—Large, showy flowers, 2–2½ inches across. Yellow ray flowers, 10–15, deeply 3-lobed at tip. Yellow disk flowers. Photos by George Clendenin, USDA-NRCS.

# Cowpen daisy *(Verbesina encelioides)*
## ASTERACEAE — SUNFLOWER OR DAISY FAMILY

**Characteristics:** Native; annual.

**Flower Bloom Period:** Spring–fall in response to rainfall.

**Growth Form:** Tall, weedy-looking plant from a taproot; height to 2 feet.

**Reproduction:** Seeds.

**General Description:** Cowpen daisy may have gotten its common name for growing near cow pens and horse lots. It is a hardy, showy wildflower but does not offer any grazing value to those animals. Leaves are triangular, with serrated edges, a characteristic gray-green color, and prominent veins underneath. Leaves sometimes have a musky odor when crushed, giving the plant the name "skunk daisy." Other common name: golden crownbeard.

**Livestock and Wildlife Value:** Cowpen daisy is rarely eaten by livestock or white-tailed deer. Seeds are frequently eaten by quail and other birds. Butterflies are often seen on the flowers.

**Management:** Usually found growing in disturbed areas but also found infrequently in good grass cover. The best way to minimize the populations of such nonpalatable plants is to practice good grazing management.

Flower—Distinctive orange/yellow, in contrast to bush sunflower, which has pale yellow flowers. Ray flowers total 7–15 and are 2-notched or -lobed at the tips. Do you remember any of our 3-lobed orange flowers? Photos by George Clendenin, USDA-NRCS.

Leaves are mostly opposite and narrowly lanceolate, with heavily toothed margins. The leaves and plant are covered by short bristly hairs, giving them a rough, sandpapery feel.

# Orange zexmenia *(Wedelia texana)*
(Sy = *Zexmenia texana*)
## ASTERACEAE—SUNFLOWER OR DAISY FAMILY

**Characteristics:** Native; perennial.

**Flower Bloom Period:** Spring–fall in response to rainfall.

**Growth Form:** Multistemmed scraggly forb from a branched root system, not a taproot; height to 18 inches.

**Reproduction:** Seeds.

**General Description:** This is a fairly common warm-season plant in the Hill Country as well as here in the Concho Valley. Orange zexmenia blooms several times each year in response to rainfall, ranking it with other flowering drought-hardy plants. Mature leaves and stems are covered by short bristly hairs, giving the plant a rough texture. Orange zexmenia is often confused with bush sunflower. See chapter 4, "Comparing and Contrasting Similar Plants," for a comparison of this plant to bush sunflower. Other common names: orange daisy, hairy wedelia.

**Livestock and Wildlife Value:** Grazed in moderation by sheep, goats, and white-tailed deer, especially when other, more desirable plants are in short supply.

**Management:** If heavy use is observed on this forb, it is a sign of too many animals. This forb, like many other perennial forbs, responds very favorably to prescribed burning. Likewise, selective mechanical brush control, especially of cedar, will usually cause an increase in these perennial forbs.

Photos by George Clendenin, USDA-NRCS.

# Plains zinnia *(Zinnia grandiflora)*
## ASTERACEAE — SUNFLOWER OR DAISY FAMILY

**Characteristics:** Native; perennial.

**Flower Bloom Period:** Spring.

**Growth Form:** Low, dense, mounding forb from a woody base; height 3–6 inches.

**Reproduction:** Seeds and short, underground rooting stems.

**General Description:** When flowering, this plant is easy to identify with its dense bouquet of yellow flowers. Flower heads often have only 3 broad petals (ray flowers) but sometimes have 4–6. More commonly found on shallow, barren soils. Leaves opposite, linear, and numerous. Other common name: prairie zinnia.

**Livestock and Wildlife Value:** Plains zinnia is rarely grazed by livestock or white-tailed deer. This plant provides a degree of soil protection and adds some organic enrichment to the soil.

**Management:** No specific management is needed to maintain this species and no control measures are warranted.

Photos by George Clendenin, USDA-NRCS.

# Rat-ear coldenia *(Tiquilia canescens)*
## BORAGINACEAE—BORAGE FAMILY

**Characteristics:** Native; perennial.

**Flower Bloom Period:** Summer.

**Growth Form:** Low mat of prostrate stems radiating from a central woody base and taproot; clump diameter 12–18 inches, height only 1–2 inches.

**Reproduction:** Seeds.

**General Description:** As the name indicates, the leaves resemble the furry ear of a rat. The Spanish name literally means "dog's ear," which is another popular common name. The gray color of the older leaves is the best identifying feature. Younger leaves are green. This plant grows mostly on shallow, sparsely vegetated areas. Because of its drought tolerance and low, spreading habit, this would make a nice ground-cover plant for xeric landscaping. Other common name: oreja de perro.

**Livestock and Wildlife Value:** This plant is rarely, if ever, grazed by livestock or white-tailed deer. Seeds are eaten by doves and scaled quail. The dense mat of stems and leaves provides excellent soil protection and a living mulch that conserves moisture.

**Management:** No specific management is needed to maintain this species and no control measures are warranted. Improved grazing management and increased grass cover will help reduce the abundance of this plant.

Photos by George Clendenin, USDA-NRCS.

# Pepperweed *(Lepidium virginicum)*
## BRASSICACEAE—MUSTARD FAMILY

**Characteristics:** Native; annual.

**Flower Bloom Period:** Spring.

**Growth Form:** Upright weedy forb; height 8–16 inches.

**Reproduction:** Seeds.

**General Description:** Flowers are tiny and white. This plant is most readily identified after flowering when the seedheads are present. Many small, round, flat seedpods form along the spiked seedhead. Pepperweed is so named because the leaves have a hot and spicy flavor. This plant is in the mustard family. Pepperweed tends to grow mostly in disturbed or sparsely grassed areas.

**Livestock and Wildlife Value:** Pepperweed is sometimes eaten in small amounts by sheep, goats, or white-tailed deer when the plant is young and tender. Seeds are eaten by birds, but the extent is unknown.

**Management:** This and other similar annual weeds can become so abundant that they compete with grasses and reduce forage production. Weed control can seldom be justified. The best way to limit weed problems is to keep a good cover of grass, which retards weed growth.

Photos by George Clendenin, USDA-NRCS.

# Cardinal flower *(Lobelia cardinalis)*
## CAMPANULACEAE—BLUEBELL OR BELLFLOWER FAMILY

**Characteristics:** Native; perennial.

**Flower Bloom Period:** Summer.

**Growth Form:** Upright forb; height 3–4 feet.

**Reproduction:** Seeds and short offshoots.

**General Description:** Grows along creek banks in seasonally wet riparian areas. Cardinal flower is one of the most beautiful of all wildflowers with its tall spikes of vivid red flowers. Leaves are alternately arranged along the stem.

**Livestock and Wildlife Value:** Use by livestock and white-tailed deer is unknown, but the plant has been found to be poisonous to humans and livestock. The flowers are a rich source of nectar and are reportedly pollinated by ruby-throated hummingbirds.

**Management:** Almost all poisonous plants are very distasteful to livestock and will not be eaten if other plants are available.

**Native American Uses:** The Pawnees used dried roots and flowers of cardinal flower in making a love charm, combined with 3 other plants: the seeds of wild columbine (*Aquilegia canadensis*) and desert biscuitroot (*Lomatium foeniculaceum* subsp. *daucifolium*), and the dried roots of ginseng (*Panax quinquefolius*). When you mixed these four ingredients with some red-earth paint, the object of your desire would fall hopelessly in love with you. If a few hairs from the desired woman were also stealthily obtained, with the help of a third-party friend, she would not be able to resist the attraction of the one who possessed the charm (Gilmore 1919).

**Toxic Agent:** Nicotine alkaloids (Hart et al. 2010).

**Toxic Description:** Lobelia poisoning is usually more prevalent in cattle and even occurs in Nilgai antelope in South Texas, usually after above-average rainfall during the previous fall and winter. Animals start out excited, then get depressed, eventually lying down and refusing food and water. However, in the Concho Valley, cardinal flower inhabits mostly riparian areas and does not pose a problem. See *Toxic Plants of Texas* (Hart et al. 2010) for livestock clinical signs and more detailed information.

Tumbleweed. Photo by George Clendenin, USDA-NRCS.

# Tumbleweed *(Salsola tragus)*
## CHENOPODIACEAE—GOOSEFOOT FAMILY

**Characteristics:** Introduced; annual.

**Flower Bloom Period:** Summer.

**Growth Form:** When young, an intricately branched, spiny-leaved weed; when mature, the characteristic tumbleweed 2–5 feet in diameter.

**Reproduction:** After the plant matures and dies, it breaks off from the root and tumbles across the ground distributing seeds along its path, a very efficient method of seed dispersal in windy West Texas.

**General Description:** It may seem odd that the stories of tumbleweeds from the old American West are actually based on this introduced plant from eastern Europe and Asia! It was possibly introduced into the United States in South Dakota in 1873 or 1874 in flaxseed imported from Russia. According to Tellman (1998), "The newly built railroad was an ideal vehicle for spreading tumbleweed throughout the West and tumbleweed's early distribution pattern shows it moving outward along railways and roadways." It is found in disturbed places such as around oil wells and on disturbed roadsides, abandoned cropland, or recently disturbed rangeland. The stems usually have reddish or purple streaks. Other common name: Russian thistle.

**Livestock and Wildlife Value:** When young, the plants are readily grazed by sheep, goats, cattle, and deer. As the plants mature and the spine-tipped leaves get hard, they are not consumed. Seeds are eaten by birds. A fence row of dead tumbleweeds can provide good cover for quail, other birds, and small mammals.

**Management:** Tumbleweed is a serious pest in cropland areas. On well-managed rangeland, tumbleweed is rarely seen. Soil disturbance such as mechanical brush control sometimes allows a crop of tumbleweeds to grow, but they will not persist.

**Toxic Agent:** Nitrate (Hart et al. 2010).

The generic name for tumbleweed is *Salsola*, which comes from the Latin word *salsus*, meaning "salty," alluding to its habitat selection of coastal and saline environments. It is appropriately pictured here, where nothing else is growing. Photos by George Clendenin, USDA-NRCS.

**Toxic Description:** All ruminants are susceptible to nitrate poisoning, with cattle poisoned most often. Nitrate poisoning is more likely if the plant grows in soils high in nitrogen, such as in livestock pens or fertilized areas. Plants with more than 1.0 percent nitrate are dangerous. Animals may die if they consume as little as 0.075 percent of their weight in nitrate. Animals with acute nitrate poisoning are often found dead with no previous history of illness. See *Toxic Plants of Texas* (Hart et al. 2010) for livestock clinical signs and more detailed information.

Habitat—Copperleaf is common among some rocky outcrops near the headwaters of the South Concho River. Photos by George Clendenin, USDA-NRCS.

# Lindheimer's copperleaf *(Acalypha phleoides)*
## EUPHORBIACEAE—EUPHORBIA OR SPURGE FAMILY

**Characteristics:** Native; perennial.

**Flower Bloom Period:** Spring–fall in response to rainfall.

**Growth Form:** Multistemmed forb from a large taproot; height about 12 inches.

**Reproduction:** Seeds. Male and female flowers are found on the same plant (unlike *Acalypha radians*).

**General Description:** The unusual flowers appear as long spikes on the end of the stems. The lower part of the flower stalk contains the female flowers, while the upper part contains the male flowers. Leaves are alternate. This plant prefers rocky slopes and limestone soils. Other common names: copperleaf, three-seeded mercury.

**Livestock and Wildlife Value:** Grazed by sheep, goats, and white-tailed deer. Seeds are eaten by birds.

**Management:** Like many other desirable perennial forbs, this plant is best maintained by good grazing management.

Photos by George Clendenin, USDA-NRCS.

# Hoary euphorbia *(Chamaesyce lata)*
## EUPHORBIACEAE—EUPHORBIA OR SPURGE FAMILY

**Characteristics:** Native; perennial.

**Flower Bloom Period:** Spring–fall in response to rainfall.

**Growth Form:** Most of the perennial spurges grow as a flattened mat on the ground. Stems radiate out from a central root.

**Reproduction:** Seeds.

**General Description:** There are 5 perennial species of spurge occurring in the Concho Valley; this is the most common species. All spurges have milky sap in the leaves and stem, although they are unrelated to milkweeds. Leaves are opposite. Other common names: perennial spurge, hoary spurge.

**Livestock and Wildlife Value:** All of the spurges provide exceptional seed value to doves, quail, and other birds. The seeds of spurge are high in fat and protein. The leaves are sometimes grazed by white-tailed deer, though they are not preferred. This plant is occasionally eaten by sheep and goats when other forage is unavailable. Some spurges are toxic to livestock when eaten in large amounts, but this has not been a problem in the region.

**Management:** Perennial spurges are a minor component of the vegetation on most ranches and pose no threat. Because of their value to birds, they are considered desirable and no control is warranted.

Photo by George Clendenin, USDA-NRCS.

# Grassland croton *(Croton dioicus)*
## EUPHORBIACEAE — EUPHORBIA OR SPURGE FAMILY

**Characteristics:** Native; perennial.

**Flower Bloom Period:** Summer.

**Growth Form:** Densely branched, bushy forb from a large woody base and large woody taproot; height to 12 inches.

**Reproduction:** Seeds.

**General Description:** Grassland croton is a perennial and appears to be a small shrub. It does not have reddish stems and dropping leaves like the annual one-seed croton. The mealy texture and distinctive silvery green color of the leaves are the best clues to identification.

**Livestock and Wildlife Value:** Seeds provide exceptional value to birds, including quail and doves. The leaves contain harsh oils that discourage grazing, but the plant is grazed to a limited extent by sheep, goats, and white-tailed deer.

**Management:** No specific management is recommended.

Photo by George Clendenin, USDA-NRCS.

# One-seed croton *(Croton monanthogynus)*
## EUPHORBIACEAE—EUPHORBIA OR SPURGE FAMILY

**Characteristics:** Native; annual.

**Flower Bloom Period:** Summer.

**Growth Form:** Single stalk at the base, branching several times to form a rounded top; height 6–12 inches depending on soil and rainfall.

**Reproduction:** Seeds.

**General Description:** This is the only common species of annual croton in the region. The stems are often reddish in color. This plant prefers calcareous soils. Other common names: doveweed, goat weed, tea weed.

**Livestock and Wildlife Value:** Excellent seed value for doves and quail. A good field of croton is a dove hunter's dream. Most landowners tolerate and appreciate a modest amount of croton in their pastures for the benefit to quail and doves. The plant is lightly grazed by sheep, goats, and white-tailed deer.

**Management:** One-seed croton densities can be stimulated by soil disturbance such as disking, fireguard construction, or mechanical brush control. Plant densities can be discouraged by keeping a good cover of grass.

Leaves—Aptly named, the mature leaves of leatherweed are coarse and feel rough like sandpaper.
Photos by George Clendenin, USDA-NRCS.

# Leatherweed croton *(Croton pottsii)*
## EUPHORBIACEAE — EUPHORBIA OR SPURGE FAMILY

**Characteristics:** Native; perennial.

**Flower Bloom Period:** Summer.

**Growth Form:** Forms large but sparse colonies interconnected by a common root system. Colonies may be 5–10 feet in diameter; individual plants grow to a normal height of 6–12 inches.

**Reproduction:** Seeds and rhizomes.

**General Description:** Mature leaves are thick and have a coarse texture; hence the name "leatherweed." Leaves are long and oval, with the silvery-green color of the leaves of other crotons.

**Livestock and Wildlife Value:** Like all other crotons, this provides good seed value to birds, but poor grazing for livestock and white-tailed deer.

**Management:** Leatherweed can become abundant and may interfere with grass production in some cases. The best way to reduce potential weed infestations is to practice conservative grazing management, which helps keep a dense cover of grass.

Photos by George Clendenin, USDA-NRCS.

# Low wild mercury *(Ditaxis humilis)*
## EUPHORBIACEAE—EUPHORBIA OR SPURGE FAMILY

**Characteristics:** Native; perennial.

**Flower Bloom Period:** Spring–fall in response to rainfall.

**Growth Form:** Low, sprawling forb from a large taproot. Stems lie horizontally on the ground and rarely grow upward.

**Reproduction:** Seeds.

**General Description:** One of the most nondescript forbs of the region. Flowers are completely inconspicuous. This plant can be identified by the long hairs present on the leaves and the 3 prominent veins, which are visible from the underside of the leaf. Seed capsule has 3 pronounced lobes, each containing 1 seed. This plant prefers rocky or shallow soils.

**Livestock and Wildlife Value:** Readily grazed by sheep, goats, and white-tailed deer. Seeds eaten by birds.

**Management:** This plant is often heavily grazed by sheep, goats, or white-tailed deer. If animal numbers remain too high and if desirable forbs are repeatedly grazed short, it will reduce the vigor of the plant and eventually cause its elimination from a pasture.

Photos by
George Clendenin,
USDA-NRCS.

# Snow-on-the-mountain *(Euphorbia marginata)*
EUPHORBIACEAE—EUPHORBIA OR SPURGE FAMILY

**Characteristics:** Native; annual.

**Flower Bloom Period:** Late summer–fall.

**Growth Form:** Tall, upright forb from a strong basal stem, much branched in the upper parts; height 2–3 feet.

**Reproduction:** Seeds.

**General Description:** Easily identified by the unique white margins on the upper leaves. The white coloration is not a part of the flowers. This plant is a close relative of poinsettia. It is one of the dozen or so species of annual *Euphorbia* found in the Concho Valley. All plants in the genus *Euphorbia* have milky sap in the leaves and stems.

**Livestock and Wildlife Value:** Seeds have excellent value for doves and quail. Since the seeds mature and shatter near the first of September, many doves shot during this time are full of these seeds. The plant is not grazed by livestock or wildlife. The milky sap contains compounds that are both distasteful and mildly toxic.

**Management:** This and other annual euphorbias can be encouraged by soil disturbance including disking, fireguard construction, or mechanical brush control. This is sometimes done intentionally to enhance food sources for doves and quail. Proper mineral supplementation, especially phosphorus, reduces livestock losses to this plant (Hart et al. 2010).

**Cultural Uses:** The caustic white sap of snow-on-the-mountain has been used to brand cattle in place of a hot branding iron!

**Toxic Agent:** Caustic white sap. Chemical compounds unknown (Hart et al. 2010).

**Toxic Description:** The white sap causes blistering of the skin and should be avoided. In most cases, livestock are poisoned by a caustic action that severely irritates the mouth and gastrointestinal tract. The plant rarely causes death. See *Toxic Plants of Texas* (Hart et al. 2010) for livestock clinical signs and more detailed information.

Photo by George Clendenin, USDA-NRCS.

# Knotweed leaf-flower *(Phyllanthus polygonoides)*
## EUPHORBIACEAE — EUPHORBIA OR SPURGE FAMILY

**Characteristics:** Native; perennial.

**Flower Bloom Period:** Summer.

**Growth Form:** Low forb with slender, weak stems from a central taproot; height 4–8 inches.

**Reproduction:** Seeds.

**General Description:** This valuable forb is most commonly found in limestone and rocky soils, often growing between rocks. Flowers are inconspicuous and droop downward along the stems. The leaves are alternately arranged.

**Livestock and Wildlife Value:** Relished by sheep, goats, and white-tailed deer, and sometimes grazed by cattle. The seeds may be eaten by birds, but this has not been confirmed.

**Management:** This and other desirable forbs can be maintained by careful attention to grazing management.

Flower—Male flowers appear on the stalk above female flowers on the same inflorescence. Notice the 3-lobed fruit. Photo by George Clendenin, USDA-NRCS.

# Texas stillingia *(Stillingia texana)*
## EUPHORBIACEAE—EUPHORBIA OR SPURGE FAMILY

**Characteristics:** Native; perennial.

**Flower Bloom Period:** Spring.

**Growth Form:** Robust multistemmed forb from a woody base and taproot; height 12–18 inches.

**Reproduction:** Flowers are unisexual, with both sexes in the same inflorescence (= monoecious). Reproduces by seeds.

**General Description:** Texas stillingia grows exclusively in shallow or rocky, calcareous soil. Leaves are long and narrow, with fine faint serrations along the edge. Flowers form on long stalks, followed by the formation of 3-lobed seedpods containing 3 large seeds. Other common name: Texas queen's delight.

**Livestock and Wildlife Value:** Good value of seeds for birds, but very poor value for livestock and white-tailed deer. Plant can be toxic to livestock, especially sheep.

**Management:** A pasture with a large amount of this plant indicates past overgrazing. The fix for overgrazed pastures often entails a variety of practices including rest from livestock grazing, reseeding, and in some cases, brush control.

**Toxic Agent:** Reported to contain cyanogenic glycosides, which release free cyanide in the rumen (Hart et al. 2010).

**Toxic Description:** Texas stillingia (*Stillingia texana*) does not pose a significant threat, unlike trecul stillingia (*Stillingia treculiana*). They both have very low palatability and are not consumed by livestock except in situations of extreme overgrazing or drought. In cases where animals have been poisoned, they are often found dead before clinical signs are observed. See *Toxic Plants of Texas* (Hart et al. 2010) for livestock signs and more detailed information.

Photo by George Clendenin, USDA-NRCS.

# Trecul stillingia *(Stillingia treculiana)*
## EUPHORBIACEAE—EUPHORBIA OR SPURGE FAMILY

**Characteristics:** Native; perennial.

**Flower Bloom Period:** Spring–summer.

**Growth Form:** Low, multistemmed forb from an enlarged taproot; height 6–12 inches.

**Reproduction:** Seeds.

**General Description:** Leaves have many coarse teeth. Flowers are inconspicuous. Seedpods are 3-lobed capsules, each containing 3 seeds.

**Livestock and Wildlife Value:** Good value of seeds for birds, but very poor value for livestock and white-tailed deer. Plant is toxic to livestock.

**Management:** Plant is found frequently in overgrazed and drought-stricken areas of the region. Good range management practices and grass establishment will help prevent this plant from increasing.

**Toxic Agent:** Reported to contain cyanogenic glycosides, which release free cyanide in the rumen (Hart et al. 2010).

**Toxic Description:** Trecul stillingia poses a significant toxic threat. This plant has poisoned sheep in drought conditions (when they were forced to eat it from lack of available palatable forage). In cases where animals have been poisoned, they are often found dead before clinical signs are observed. See *Toxic Plants of Texas* (Hart et al. 2010) for livestock clinical signs and more detailed information.

Many plants have hairs, but not all hairs are "stinging hairs." Supposedly, only 4 plant families have stinging hairs: Euphorbiaceae, Hydrophyllaceae, Loasaceae, and Urticaceae. All these families have representative species in West Texas. Each stinging mechanism is varied and complex. Current studies suggest that *Tragia ramosa* has a single stinging cell and 3 lateral cells. The stinging cell has protein-like substances that are released when it touches our skin, causing a contact dermatitis reaction. Photos by George Clendenin, USDA-NRCS.

# Noseburn *(Tragia ramosa)*
## EUPHORBIACEAE—EUPHORBIA OR SPURGE FAMILY

**Characteristics:** Native; perennial.

**Flower Bloom Period:** Spring–fall in response to rainfall.

**Growth Form:** Low, weak-stemmed forb from a central taproot.

**Reproduction:** Seeds.

**General Description:** One of the few plants in the region that produce a stinging skin irritation. The plant can be obscure until you learn to recognize it; you will often feel the sting before you see the plant. Leaves, alternately arranged, are triangular and have prominent serrations on the edge. Long stinging hairs can be easily seen on leaves and stems. Flowers are inconspicuous; seedpod is a 3-lobed capsule. Other common name: stinging nettle.

**Livestock and Wildlife Value:** Oddly enough, the plant is grazed by sheep, goats, and white-tailed deer and is eaten to a greater extent during drought when other plants are less available. Seeds are eaten by quail and other birds.

**Management:** No management is needed for this plant.

Leaflets—The leaflets of peavine are variable and will often, but not always, have a blunt tip with a notch. Leaves are odd-pinnately compound. Photos by George Clendenin, USDA-NRCS.

# Nuttall peavine *(Astragalus nuttallianus)*
## FABACEAE—LEGUME OR BEAN FAMILY

**Characteristics:** Native; annual.

**Flower Bloom Period:** Early spring.

**Growth Form:** Low annual forb, with stems often trailing across the ground; normal height only 3–6 inches.

**Reproduction:** Seeds.

**General Description:** Nuttall peavine is a cool-season plant and not a true vetch, though it is often confused with deer pea vetch (*Vicia ludoviciana*). Both may be seen growing side by side during spring. Nuttall peavine lacks the forked tendrils of vetch. The seedpods of Nuttall peavine are strongly curved and have a pronounced groove along the lower edge. Other common names: peavine, Nuttall milkvetch.

**Livestock and Wildlife Value:** A paradoxical plant that can be both valuable and toxic to livestock. On limestone soils, the plant can be toxic to cattle, sheep, and goats, especially if animals are disturbed or stressed. Ranchers who are aware that peavine is being eaten should be careful to avoid the stress of working or moving livestock. If this precaution is followed, peavine usually causes no problems. Peavine is a high-quality spring forage for livestock and white-tailed deer. The seeds are frequently eaten by quail and other birds.

**Management:** Like other winter-annual forbs, peavine depends on good fall or winter rains for germination and growth. In a dry fall and winter, there will be almost no peavine; in a wet winter, there will often be a profusion of peavine and other winter annuals. Nuttall peavine, like all legumes, is capable of adding nitrogen enrichment to the soil.

Photos by
George Clendenin,
USDA-NRCS.

# Purple dalea *(Dalea lasiathera)*
## FABACEAE—LEGUME OR BEAN FAMILY

**Characteristics:** Native; perennial.

**Flower Bloom Period:** Spring.

**Growth Form:** Prostrate forb from a large central taproot. Stems radiate 8–12 inches from the center, but to a height of only 2–4 inches.

**Reproduction:** Seeds.

**General Description:** The attractive purple flowers are the best clue for identification; without flowers, the plant is best identified by smell. The crushed leaves have a very strong and peculiar odor. Leaves are naked, lacking any hairs. Purple dalea, like other legumes, can fix nitrogen into the soil.

**Livestock and Wildlife Value:** Purple dalea is eaten by sheep, goats, and white-tailed deer, although it is not highly preferred.

**Management:** Conservative grazing management and periodic rest are the best ways to ensure the continuation and increase of desirable perennial forbs.

Photos by George Clendenin, USDA-NRCS.

# Dwarf dalea *(Dalea nana)*
## FABACEAE—LEGUME OR BEAN FAMILY

**Characteristics:** Native; perennial.

**Flower Bloom Period:** Spring–early summer.

**Growth Form:** Short, upright forb from a large taproot; height 4–8 inches.

**Reproduction:** Seeds.

**General Description:** Small yellow flowers are clustered in a feathery spike. Usually seen without the flowering stalk or seedhead, in which case the plant is rather nondescript. Identification as a dalea can be confirmed by the characteristic odor of the crushed leaves. Leaves are densely covered by hairs.

**Livestock and Wildlife Value:** Plant is eaten by sheep, goats, and white-tailed deer.

**Management:** Grazing management is the factor that most affects the abundance of this and most other desirable perennial forbs.

Photos by George Clendenin, USDA-NRCS.

# Velvet bundleflower *(Desmanthus velutinus)*
## FABACEAE—LEGUME OR BEAN FAMILY

**Characteristics:** Native; perennial.

**Flower Bloom Period:** Spring–fall in response to rainfall.

**Growth Form:** Moderate-sized forb from a large woody taproot; height will vary from 6 to 12 inches depending on soil, rainfall, and vigor of plant.

**Reproduction:** Seeds.

**General Description:** Almost all legumes have compound leaves that are composed of several to numerous leaflets. This plant has double-compound leaves (bipinnately compound). Leaves and stems are often covered by a velvety layer of fine hairs. Flowers are unimpressive, lacking petals. Seedpods are very distinctive. Each pod is in a cluster composed of several long beans that are bundled together at the base. Each bean has several seeds. This plant is more often found in shallow, rocky soil. Like all legumes, this plant has the ability to convert atmospheric nitrogen to soil nitrogen by the activity of rhizobium bacteria that live in the root system.

**Livestock and Wildlife Value:** Very desirable food for sheep, goats, and deer. Seeds are frequently eaten by quail and other birds.

**Management:** This plant is one of the most preferred forbs and is very susceptible to overuse if livestock or white-tailed deer numbers are too high. Careful and consistent attention to grazing management and deer numbers is necessary in order to maintain the abundance and vigor of this plant. This plant can be used as an indicator of grazing management. If plants are abundant, large, and leafy, it is a positive sign of good management. If plants are scarce, small, stunted, or confined to protected areas, it is a sign that animal numbers need to be reduced and a system of pasture rotation implemented.

Photos by George Clendenin, USDA-NRCS.

234

# Texas bluebonnet *(Lupinus subcarnosus)*
## FABACEAE—LEGUME OR BEAN FAMILY

**Characteristics:** Native; annual.

**Flower Bloom Period:** Early spring.

**Growth Form:** Upright forb from a taproot; height to about 12 inches.

**Reproduction:** Seeds are formed in large pods. As the pods mature and dry, they forcefully split apart in 2 spiraling halves, throwing the seeds some distance. The seeds are very attractive upon close examination, with intricate patterns of several colors.

**General Description:** Bluebonnet needs no description to the citizens of Texas. It is one of the most beloved of all spring wildflowers. In 1901 the Texas legislature adopted *Lupinous subcarnosus* as the State Flower of Texas. However, there are 6 bluebonnet species, which caused great confusion as to which one was actually intended to be the state flower. After 70 years of argument, in 1971, the Texas legislature adopted all 6 species of bluebonnet as the State Flower of Texas. So in essence Texas really has 6 official state flowers, all of them bluebonnets!

**Livestock and Wildlife Value:** Not considered good forage for livestock or white-tailed deer, but sometimes consumed in small to moderate amounts. Seeds are eaten by birds. The nitrogen fixation by the roots of this plant provides indirect benefit to ranchers.

**Management:** In the Concho Valley, unlike in the Hill Country, bluebonnet is not a normal wildflower each spring. Bluebonnets are abundant only once every 5 to 10 years when rainfall in fall and winter is above average. Pastures may be devoid of bluebonnets for many years, and then when conditions are right, they can be abundant. This is the nature of annual forbs, with their normal boom-and-bust cycle of abundance. For those who wish to do so, the seeds of bluebonnet can be purchased to plant into desired wildflower areas. Bluebonnet seeds should be planted in September or October and be scarified for best germination.

Photos by George Clendenin, USDA-NRCS.

# Bur clover *(Medicago minima)*
## FABACEAE—LEGUME OR BEAN FAMILY

**Characteristics:** Introduced; annual.

**Flower Bloom Period:** Spring.

**Growth Form:** Low, crawling annual from a central taproot; diameter 8–24 inches, height 3–6 inches depending on rainfall and soil.

**Reproduction:** Seeds are formed in spiral pods. Each pod has numerous hooked bristles or "burs" that attach themselves to passing animals, thus spreading the pod and seeds.

**General Description:** The characteristic clover leaf has 3 leaflets (pinnately compound), each with finely toothed edges. Flowers are tiny and yellow. Botanically, this plant is not a true clover. It is a medic and is closely related to alfalfa. Bur clover was introduced from Eurasia. Like all other legumes, it provides nitrogen enrichment of soil.

**Livestock and Wildlife Value:** Traditional sheep and goat ranchers have a strong dislike for bur clover if they are producing wool or mohair. The burs get so entangled into hair coats that the value of the fleece and hair is greatly diminished. Otherwise, bur clover is a valuable forage plant for sheep, goats, cattle, and white-tailed deer.

**Management:** Like most other annual forbs, this plant is more commonly found in bare or sparse areas and is not common in well-grassed areas. A variety of bur clover is now commercially available for those who wish to plant it in food plots or use it in range planting mixtures.

How do the leaves close by themselves? The leaves are touch sensitive and rapidly close after touching (photo on right). This folding of leaves is a result of pressure changes in the joints of the leaflet and the leaf. This is thought to be a self-defense mechanism against herbivores or an adaptation to reduce water loss within the plant. Photos by George Clendenin, USDA-NRCS.

# Sensitive briar *(Mimosa nuttallii)*
## FABACEAE—LEGUME OR BEAN FAMILY

**Characteristics:** Native; perennial.

**Flower Bloom Period:** Spring–fall in response to rainfall.

**Growth Form:** Long, prostrate stems from a large central taproot. Stems may be 3 or 4 feet long but are only several inches high.

**Reproduction:** Seeds.

**General Description:** This plant is familiar to all kids raised in the country. The leaves are sensitive and quickly fold up with even the slightest touch. The long stems, the leaf stalks, and the pods are all covered with thousands of sharp, short, recurved spines. Flowers are very attractive pink puffballs, which are composed of thousands of stamens. Leaves are bipinnately compound. The large, deep taproot of this and most other perennial legumes provides indirect value to the soil, penetrating deep subsoil layers and allowing the movement of water and air.

**Livestock and Wildlife Value:** Leaves and stems are eaten by sheep, goats, white-tailed deer, and sometimes cattle, especially when the plant is young and the spines are still soft. Seeds are eaten by birds.

**Management:** Desirable perennial forbs are best managed by careful grazing management and adjustments in the deer population.

Leaves—The leaves of Texas snoutbean are trifoliate and have a strong keel-shaped midrib and prominent veins, which are typical of the genus *Rhyncosia* (from the Greek *rhynchos,* meaning "snout" or "beak"). Photos by George Clendenin, USDA-NRCS.

# Texas snoutbean *(Rhynchosia senna)*
## FABACEAE—LEGUME OR BEAN FAMILY

**Characteristics:** Native; perennial.

**Flower Bloom Period:** Spring–fall in response to rainfall.

**Growth Form:** Large taprooted forb with long, weak, vining stems that grow up into supporting vegetation or across the ground. Plant can be up to 4 feet tall and 4 feet across if there is a shrub to support it.

**Reproduction:** Seeds.

**General Description:** Leaves have a unique pattern of veins that give the plant a textured appearance unlike that of any other plant in the region. Leaves are composed of 3 leaflets. Flowers are small and yellow.

**Livestock and Wildlife Value:** Texas snoutbean provides good forage for sheep, goats, and white-tailed deer. Birds presumably eat the seeds, but this is not documented.

**Management:** Conservative grazing management is important for the perpetuation of this plant.

Photos by George Clendenin, USDA-NRCS.

# Twin-leaf senna *(Senna roemeriana)*
## FABACEAE—LEGUME OR BEAN FAMILY

**Characteristics:** Native; perennial.

**Flower Bloom Period:** Spring–fall in response to rainfall.

**Growth Form:** Stout, upright, multistemmed forb, from a very large woody taproot; height 12–18 inches.

**Reproduction:** Seeds.

**General Description:** As the name indicates, each leaf is composed of twin leaflets, somewhat resembling a pair of rabbit ears. Flowers are showy, with 5 yellow petals.

**Livestock and Wildlife Value:** Plant is very toxic to cattle and goats. Sheep are more resistant but can be poisoned if large amounts are eaten. Other common name: two-leaf senna.

**Management:** The best management to minimize losses from toxic plants is to maintain good range condition and forage availability. This will allow animals to graze naturally and avoid eating toxic and distasteful plants. Supplemental feed, especially with phosphorous, may also reduce livestock losses due to twin-leaf senna. Pastures that are dominated by abnormally large numbers of toxic plants need aggressive range restoration methods, which may involve reseeding with desirable native grasses. Small populations can be treated with chemical herbicides.

**Toxic Agent:** Unknown.

**Toxic Description:** Twin-leaf senna is toxic to cattle, goats, and horses. Sheep are more resistant but can be poisoned if they eat too much. A lethal dose for cattle and goats is about 1 percent of their body weight eaten for 5 to 10 days. Young kid goats may be poisoned with much less fresh plant material. In goats, animals die suddenly of heart failure. In cattle, large skeletal muscles are most affected. Poisoned horses and sheep die of liver failure. See *Toxic Plants of Texas* (Hart et al. 2010) for livestock clinical signs and more detailed information.

Photos by
George Clendenin,
USDA-NRCS.

# Deer pea vetch *(Vicia ludoviciana)*
## FABACEAE — LEGUME OR BEAN FAMILY

**Characteristics:** Native; annual.

**Flower Bloom Period:** Mid to late spring.

**Growth Form:** Viny, spreading forb that grows upon other plants for support.

**Reproduction:** Seeds are formed in pods. When the pods dry, they split forcefully and throw the seeds some distance from the plant.

**General Description:** This plant may be confused with Nuttall peavine. Vetch has forked tendrils at the tip of each stem that wrap around nearby vegetation for support. Peavine lacks these tendrils. The seedpods of vetch are flattened when young but get round at maturity. The seedpods of peavine have a conspicuous groove running along the lower edge; this groove is absent in vetch. The leaves are pinnately compound.

**Livestock and Wildlife Value:** Leaves are relished by livestock and white-tailed deer. Seeds are eaten by quail and other birds. Vetch and other legumes add nitrogen to the soil, which benefits other plants and indirectly benefits livestock and wildlife.

**Management:** Vetch may be very abundant in wet years and totally absent in normal or dry years. Management has little effect on the abundance of vetch; however, extreme heavy grazing will reduce if not eliminate successful flowering and seed production, which will limit future establishment.

Photos by George Clendenin, USDA-NRCS.

# Mountain pink *(Centaurium beyrichii)*
## GENTIANACEAE — GENTIAN FAMILY

**Characteristics:** Native; annual.

**Flower Bloom Period:** Late spring–early summer.

**Growth Form:** Grows as a small, compact bouquet of pink flowers from a central taproot, height about 6–12 inches.

**Reproduction:** Seeds.

**General Description:** An unmistakable and beautiful plant when flowering. Usually grows in dense clumps shaped like an inverted cone, giving the appearance of a pink bouquet of flowers. All the leaves in mountain pink are linear. In the Concho Valley and west central Texas, *C. beyrichii* (mountain pink) grows on limestone hills on shallow soils. Elsewhere, *C. calycosum* (rosita) grows on moist soils.

**Livestock and Wildlife Value:** There is some evidence that this plant is toxic to livestock, although detailed information is lacking. There is no known value to wildlife.

**Management:** Maintaining good grass cover with good grazing management is the best way to minimize the establishment of this plant in pastures.

**Toxic agent:** Unknown.

**Toxic description:** Suspected to be poisonous to cattle, sheep, and goats. The toxic dose is estimated to be between 0.5 and 1.0 percent of an animal's body weight, consumed daily for several days. See *Toxic Plants of Texas* (Hart et al. 2010) for livestock clinical signs and more detailed information.

Photos by George Clendenin, USDA-NRCS.

# California filaree *(Erodium cicutarium)*
## GERANIACEAE — GERANIUM FAMILY

**Characteristics:** Introduced; annual or biennial.

**Flower Bloom Period:** Later winter–early spring.

**Growth Form:** Low-growing forb from a central taproot; height 3–6 inches.

**Reproduction:** Seeds, which are attached to awns, plant themselves in a similar fashion as those of Texas filaree.

**General Description:** Leaves are much dissected and resemble the leaves of a fern. Flowers are small and pink. The "stork's bill" fruits are smaller and shorter than those of Texas filaree. This plant is native to Europe.

**Livestock and Wildlife Value:** Like its cousin, Texas filaree, this plant provides good seasonal grazing for livestock and deer.

**Management:** An abundance of winter weeds such as filaree is a mixed blessing to ranchers. On one hand, the flush of nutritious forage is desirable; on the other hand, these plants also suck moisture out of the soil and may materially reduce spring grass production. A moderate amount of these winter annuals is usually considered desirable to livestock ranching and wildlife. If the density of these plants becomes extreme, it usually indicates a lack of good grass cover.

Photos by George Clendenin, USDA-NRCS.

# Texas filaree *(Erodium texanum)*
## GERANIACEAE — GERANIUM FAMILY

**Characteristics:** Native; annual or biennial.

**Flower Bloom Period:** Spring.

**Growth Form:** Low-growing forb from a taproot; height 3–6 inches.

**Reproduction:** Seeds have a peculiar means of planting themselves. Each seed is shaped like a spearhead and is attached to a long awn. The awn is "hygroscopic," meaning that it twists and untwists in response to moisture and humidity. The dry awn will corkscrew in a tight spiral. As the awn absorbs moisture, it uncurls. This action helps to plant the sharp seed into the ground. Other common name: stork's bill.

**General Description:** In typical fashion, this winter forb usually germinates in the fall and grows as a "winter rosette" of leaves until flowering. This rosette, or cluster of leaves, lies flat on the ground during early winter, which gives the plant some initial resistance to grazing. As the plant matures, leaves begin to grow upward. Leaves are broad and palmately lobed. Leaves or leaf veins often turn red, giving a striking and attractive appearance. Flowers are large and purple and resemble the flowers of geranium. The seeds are formed in clusters of 5. As the pods mature, the individual awns and seeds separate from each other. The green seedpods give this plant the name "stork's bill."

**Livestock and Wildlife Value:** This plant is highly valued by ranchers who know that all classes of livestock and white-tailed deer readily graze filaree in the spring. Seeds are occasionally eaten by quail.

**Management:** An abundance of winter weeds such as filaree is a mixed blessing to ranchers. On one hand, the flush of nutritious forage is desirable; on the other hand, these plants also suck moisture out of the soil and may materially reduce spring grass production. A moderate amount of these winter annuals is usually considered desirable to livestock ranching and wildlife. If the density of these plants becomes extreme, it usually indicates a lack of good grass cover.

Photos by George Clendenin, USDA-NRCS.

# Trailing ratany *(Krameria lanceolata)*
## KRAMERIACEAE—RATANY FAMILY

**Characteristics:** Native; perennial.

**Flower Bloom Period:** Late spring–early summer.

**Growth Form:** Long, trailing stems from a large, central, woody taproot; diameter 3–4 feet, height only several inches.

**Reproduction:** Seeds are produced inside a large bur with sharp spines. The bur is transported by passing animals.

**General Description:** This plant is supported by a very large, deep taproot. The taproot penetrates deeply and grows into cracks in the bedrock. This immense root system provides large energy reserves and extreme drought hardiness. The stems are reddish and covered by fine fuzz. Flowers are distinctive in color and shape, as shown in the photos. The simple leaves have smooth margins and are alternately arranged along the stem.

**Livestock and Wildlife Value:** Trailing ratany is readily grazed by sheep, goats, and deer.

**Management:** This and other desirable perennial forbs are best maintained by good grazing management.

**Minty Fresh**—If you find yourself smelling a strong minty odor after walking or riding through a ranch pasture in the Concho Valley, chances are you are smelling this plant. Photos by George Clendenin, USDA-NRCS.

# False pennyroyal *(Hedeoma drummondii)*
## LAMIACEAE — MINT FAMILY

**Characteristics:** Native; perennial.

**Flower Bloom Period:** Spring.

**Growth Form:** Bushy to sprawling forb; height 6–12 inches in this region.

**Reproduction:** Seeds.

**General Description:** The best way to verify the identification of this plant is by smelling the crushed leaves. The smell is strong, minty, and refreshing. Flowers are small and inconspicuous. All plants in the mint family have square stems, and leaves that occur in opposite pairs along the stem. Pennyroyal has mostly linear leaves. Other common names: Drummond's false pennyroyal, mock pennyroyal.

**Livestock and Wildlife Value:** Infrequently grazed by sheep, goats, and white-tailed deer.

**Management:** No specific management is needed to maintain this plant.

Photos by George Clendenin, USDA-NRCS.

# Common horehound *(Marrubium vulgare)*
## LAMIACEAE — MINT FAMILY

**Characteristics:** Introduced; perennial.

**Flower Bloom Period:** Spring.

**Growth Form:** Robust multistemmed forb; height 12–18 inches.

**Reproduction:** Seeds.

**General Description:** Plants in the Lamiaceae (mint) family can be recognized in the field by their square stems and opposite leaves. The most characteristic trait of horehound is the strongly wrinkled texture of the leaves. Leaves and stems are covered with a fine layer of dense fuzz, sometimes giving the plant a silvery white appearance. Flowers are inconspicuous, occurring in spikes. Seeds are formed in bur-like capsules covered by bristles. Other common name: white horehound.

**Livestock and Wildlife Value:** This plant is notoriously disliked by ranchers producing wool and mohair. The burs, covered by bristles, frequently get compacted in the hair coat, greatly reducing the value of the fleece and hair. The leaves are sometimes grazed by sheep, goats, and white-tailed deer, especially in a dry winter when no other green forage is available.

**Management:** This plant is most commonly found near chronically overgrazed watering locations, pens, bedding areas, and other places where sheep have historically congregated. The plant thrives best in sparsely vegetated areas and is not commonly found in good grass cover. Populations can persist many years after sheep have been removed from the range, and herbicidal spot control in late winter or early spring may be needed to control infestations of this exotic weed.

Photos by George Clendenin, USDA-NRCS.

# Lemon beebalm *(Monarda citriodora)*
## LAMIACEAE — MINT FAMILY

**Characteristics:** Native; annual.

**Flower Bloom Period:** Spring.

**Growth Form:** Tall, upright forb; height 12–24 inches depending on rainfall and soil.

**Reproduction:** Seeds.

**General Description:** Lemon beebalm is native to Texas. Like all other plants in the mint family, lemon beebalm has opposite leaf arrangement; leaves always occur as matching pairs along the stem. Flower heads are large and showy. Other common names: beebalm, lemon mint, horse mint.

**Livestock and Wildlife Value:** Occasionally grazed by livestock or white-tailed deer. The seeds may have some value to birds, but this has not been documented.

**Management:** This popular wildflower is included in many wildflower seed mixes. Ranchers do not normally encourage the plant but appreciate it in small amounts for its attractive flowers. Large, thick areas of beebalm may reflect a lack of good grass cover and indicate the need for improved grazing management.

**Other Uses:** Lemon beebalm contains citronellol, a fragrant chemical that occurs naturally in more than 30 plant oils as well as in black tea, wine, and certain fruits. Citronellol attracts mites and is therefore used, along with other substances, in pesticides. The first pesticide with citronellol was registered in 2005. It has not been shown to be harmful to humans in small amounts, as it has also has a long history of use in cosmetics, flavorings, and fragrances (Diggs, Lipscomb, and O'Kennon 1999; US EPA 2005).

Photos by George Clendenin, USDA-NRCS.

# Mealycup sage *(Salvia farinacea)*
## LAMIACEAE — MINT FAMILY

**Characteristics:** Native; perennial.

**Flower Bloom Period:** Spring–early summer.

**Growth Form:** Tall forb from a large woody taproot; height to 24 inches when flowering.

**Reproduction:** Seeds.

**General Description:** It is a nice surprise to see this plant in a summer pasture, as its violet-purple blooms last almost all summer long. It is also one of my personal favorites as a hardy, drought-tolerant landscape plant around the house. Flowers form in long spikes at the top of long waving stems. The entire plant is covered by a fine dense layer of short fuzzy hair, giving the plant a powdery or mealy appearance. Leaves are lanceolate to oblong-lanceolate. Other common name: mealy sage, mealy blue sage.

**Livestock and Wildlife Value:** Rarely if ever grazed by livestock or white-tailed deer.

**Management:** This plant sometimes indicates past overgrazing since it can thrive on poor, eroded soils and is extremely unpalatable to livestock. It is widely used in wildflower gardens and native landscapes. Several "improved" varieties have been developed for the ornamental landscape business.

Photo by George Clendenin, USDA-NRCS.

# Lance-leaf sage *(Salvia reflexa)*
## LAMIACEAE — MINT FAMILY

**Characteristics:** Native; annual.

**Flower Bloom Period:** Spring.

**Growth Form:** Upright forb; height 12–18 inches.

**Reproduction:** Seeds.

**General Description:** Leaves are long and narrow (lance-leaf), botanically described as lanceolate to linear-lanceolate, arranged in opposite pairs along the square stem. Flowers are unremarkable. All plants in this group (*Salvia*) are strongly aromatic. Other common name: Rocky Mountain sage.

**Livestock and Wildlife Value:** Rarely if ever grazed by livestock or wildlife.

**Management:** This weedy plant is found most often in disturbed locations and is most abundant in seasonally moist areas. Look for this plant in hay fields before mowing. See toxicity information. A good cover of grass is the best way to ensure that this plant does not become a weed problem.

**Toxic Agent:** Unknown.

**Toxic Description:** There are no documented reports of this plant causing poisoning under range or pasture conditions, although it has been suspected. However, reported cases of poisonings in the United States are limited to cattle and horses that consumed contaminated hay. Experimental feeding trials have shown that sheep are also susceptible. It is not known how much plant material must be eaten to cause toxicity. See *Toxic Plants of Texas* (Hart et al. 2010) for livestock clinical signs and more detailed information.

Texas salvia is shown here growing with bush sunflower. Photos by George Clendenin, USDA-NRCS.

# Texas salvia *(Salvia texana)*
## LAMIACEAE — MINT FAMILY

**Characteristics:** Native; perennial.

**Flower Bloom Period:** Spring.

**Growth Form:** Stout-stemmed, upright forb from a woody taproot; height 8–12 inches.

**Reproduction:** Seeds.

**General Description:** Texas salvia is best identified by the extremely bristly-hairy texture of the stems and leaves. Both Texas salvia and skullcap have violet-purple flowers with two prominent white markings toward the base. However, closer observation reveals that skullcap often has spots on these white markings. Skullcap also lacks the bristly appearance of Texas salvia and has different seedpods. Leaves of Texas salvia are lanceolate. This plant grows best in very shallow or rocky soil. Other common name: Texas sage.

**Livestock and Wildlife Value:** Sometimes eaten by sheep and white-tailed deer, but not considered a desirable forage plant.

**Management:** No particular management is recommended for this plant.

Flower—Both Texas salvia and skullcap have violet-purple flowers with two prominent white markings toward the base. However, skullcap often has spots on these white markings. Photos by George Clendenin, USDA-NRCS.

# Skullcap *(Scutellaria drummondii)*
## LAMIACEAE — MINT FAMILY

**Characteristics:** Native; annual or short-lived perennial.

**Flower Bloom Period:** Spring.

**Growth Form:** Low, bushy forb; height 6–8 inches.

**Reproduction:** Seeds.

**General Description:** Skullcap can sometimes be confused with pennyroyal when flowers are not present, and it can also be confused with Texas salvia when flowers are present. Skullcap lacks the strong minty odor of pennyroyal. The seedpods also differ: those of skullcap are small, rounded capsules with a flat or concave top. A peculiar method of seed dispersal takes place as raindrops hit the mature capsule, causing it to explode and throwing the seeds some distance. Other common name: Drummond's skullcap.

**Livestock and Wildlife Value:** Occasionally grazed by sheep, goats, or white-tailed deer, but not considered an important forage plant. The seeds are likely eaten by some kinds of grassland birds.

**Management:** No particular management is recommended for this species.

You will know if you have spent any time in contact with this plant, as your pant legs will be covered by the bristly leaves! Photos by George Clendenin, USDA-NRCS.

# Stickleaf *(Mentzelia oligosperma)*
## LOASACEAE—STICKLEAF OR BLAZINGSTAR FAMILY

**Characteristics:** Native; perennial.

**Flower Bloom Period:** Spring–fall in response to rainfall.

**Growth Form:** Bushy, sprawling, much-branched forb from a large, rounded tuber.

**Reproduction:** Seeds are spread when seedpods catch in the hair coat of passing animals.

**General Description:** Frequently found growing in the protection of prickly pear or other spiny plants, indicating its value to grazing animals. The entire plant is densely covered by short hooked bristles that do not sting, giving the plant a scratchy texture. The hooked bristles very easily get attached to clothing or to hair coats of animals and thus can be a problem for sheep producers. Star-shaped flowers are small but very attractive, with long filaments coming from the center. Other common name: chicken thief.

**Livestock and Wildlife Value:** Commonly grazed by sheep, goats, and white-tailed deer; sometimes grazed by cattle.

**Management:** This desirable forb is maintained by conservative grazing management.

Photo by Gary Rea, USDA-NRCS.

Photos by George Clendenin, USDA-NRCS.

# Indian mallow *(Abutilon fruticosum)*
## MALVACEAE — MALLOW FAMILY

**Characteristics:** Native; perennial.

**Flower Bloom Period:** Spring–fall in response to rainfall.

**Growth Form:** Tall, rigid, upright forb from large taproot.

**Reproduction:** Seeds.

**General Description:** The leaves are heart shaped and covered by a dense layer of fine hair, giving the plant a grayish-green color. The small flowers are orange to yellow and resemble miniature hibiscus flowers. This plant grows most commonly in shallow, rocky soil. Seedpod is a capsule with several compartments, each containing 2 or more seeds. Other common name: pelotazo.

**Livestock and Wildlife Value:** Grazed by sheep, goats, and white-tailed deer. The seed capsules are commonly eaten by turkeys, while the individual seeds are sometimes eaten by quail and other songbirds.

**Management:** This plant is rarely overgrazed except in extreme cases. It can be maintained with average grazing management or increased with good grazing management.

Leaves—The leaves of wine cup are palmately lobed.

Community—This beautiful landscape shows a community of wine cup in the foreground with the pink showy evening primrose in the background. Photos by George Clendenin, USDA-NRCS.

# Wine cup *(Callirhoe involucrata)*
## MALVACEAE — MALLOW FAMILY

**Characteristics:** Native; perennial.

**Flower Bloom Period:** Late winter–spring.

**Growth Form:** Low, spreading forb from an enlarged taproot. Clumps grow to a diameter of 18–36 inches and a height of about 12 inches. Usually much smaller in grazed pastures.

**Reproduction:** Seeds.

**General Description:** A very attractive spring wildflower with masses of large purple flowers. Flowers have 5 petals that are initially folded upward in the shape of a wine cup. As flowers age, the petals open up and spread out. Flower stalks are covered with long hairs. Other common names: low poppy-mallow, purple poppy-mallow.

**Livestock and Wildlife Value:** Very palatable and readily grazed by sheep, goats, white-tailed deer, and cattle. Seeds are sometimes eaten by birds.

**Management:** Palatable forbs that are preferred by livestock and deer require very good grazing management and periodic rest from grazing. White-tailed deer must be maintained at a low to moderate population.

**Native American Uses:** The Oglalas and Teton Dakotas used wine cup for "smoke treatment medicine." The root was also boiled for other medicinal uses (Gilmore 1919).

Photos by George Clendenin, USDA-NRCS.

# Bladderpod sida *(Rhynchosida physocalyx)*
MALVACEAE—MALLOW FAMILY

**Characteristics:** Native; perennial.

**Flower Bloom Period:** Spring–summer in response to rainfall.

**Growth Form:** Weak-stemmed, prostrate forb from a taproot.

**Reproduction:** Seeds.

**General Description:** The stems of this plant grow horizontally across the ground. Leaves and stems are covered by dense hairs. If the hairs are examined closely with magnification, they reveal an interesting starburst pattern. The creamy yellow flowers have 5 broad, overlapping petals that are equal in length to the 5 green sepals. Each flower and fruit is contained within a large, inflated, 5-ribbed calyx (= sepals collectively); hence the name bladderpod. The pod turns black at maturity and contains about a dozen black seeds. Other common name: buffpetal.

**Livestock and Wildlife Value:** Frequently eaten by sheep, goats, and white-tailed deer during spring and summer.

**Management:** Conservative grazing management and control of white-tailed deer numbers will help maintain this desirable forb in pastures.

Photo by Steve Nelle, USDA-NRCS.

Photo by George Clendenin, USDA-NRCS.

# Spreading sida *(Sida abutifolia)*
## MALVACEAE — MALLOW FAMILY

**Characteristics:** Native; perennial.

**Flower Bloom Period:** Spring–summer in response to rainfall.

**Growth Form:** Small weak-stemmed forb, spreads from a small taproot.

**Reproduction:** Seeds.

**General Description:** Leaves are small, with blunt, rounded serrations. Flowers are ½–¾ inch in diameter, with 5 yellow petals. Stems are weak and covered by hairs.

**Livestock and Wildlife Value:** Commonly eaten by sheep, goats, and white-tailed deer. This is often the most common perennial forb observed on ranches in midsummer, and it can withstand hot, dry conditions better than many other forbs. Because of its abundance, it can make up an important part of animal diets during summer. Seedpods and seeds are sometimes eaten by birds.

**Management:** Ongoing grazing management and attention to deer numbers will help maintain this desirable forb.

Photos by George Clendenin, USDA-NRCS.

# Narrowleaf globemallow
*(Sphaeralcea angustifolia)*
**MALVACEAE — MALLOW FAMILY**

**Characteristics:** Native; perennial.

**Flower Bloom Period:** Spring–fall in response to rainfall.

**Growth Form:** Tall, upright forb from a large woody root system, sometimes forms loose colonies; height 18–30 inches.

**Reproduction:** Seeds and short, branching, underground stems.

**General Description:** Leaves are long and narrow, toothed but unlobed, and covered by a dense layer of hairs, giving them a soft texture. Flowers are usually a soft salmon pink.

**Livestock and Wildlife Value:** Readily grazed by livestock and white-tailed deer. Seedpods may be eaten by turkeys.

**Management:** Conservative grazing management will help perpetuate or increase the abundance of this plant.

Leaf—All leaves in the genus *Sphaeralcea* have pubescence (fine short hairs) that under magnification appear to radiate like a starburst (stellate). Scarlet globemallow leaves are dissected into 3 main lobes and several secondary lobes, unlike the narrow, unlobed leaves of narrowleaf globemallow. Photos by George Clendenin, USDA-NRCS.

# Scarlet globemallow *(Sphaeralcea coccinea)*
## MALVACEAE—MALLOW FAMILY

**Characteristics:** Native; perennial.

**Flower Bloom Period:** Spring–early summer.

**Growth Form:** Short, upright forb from a large branching root system, may form colonies; height 8–12 inches.

**Reproduction:** Seeds and rhizomes.

**General Description:** This species of globemallow grows only in the far western part of the Concho Valley. Flowers are bright scarlet orange. Other common names: caliche globemallow, red false mallow.

**Livestock and Wildlife Value:** Commonly grazed by livestock and white-tailed deer.

**Management:** This and all other palatable grazing forbs depend on good grazing management to maintain good health and productivity.

**Native American Uses:** The Dakotas utilized this plant by "chewing it to a paste, which was rubbed over the hands and arms, thus making them immune to the effect of scalding water, so . . . these men were able to take up hot pieces of meat out of the kettle over the fire. The plant was also chewed and applied to inflamed sores and wounds as a cooling and healing salve" (Gilmore 1919). The Cheyennes also knew this plant by a name translated as "sweet medicine" and used it for ceremonial and medicinal purposes (Ajilvsgi 2003).

Flower—Angel trumpets have some of the longest flowers found in Texas plants. Photos by George Clendenin, USDA-NRCS.

# Angel trumpets *(Acleisanthes longiflora)*
NYCTAGINACEAE—FOUR-O'CLOCK FAMILY

**Characteristics:** Native; perennial.

**Flower Bloom Period:** Spring–fall in response to rainfall.

**Growth Form:** Sprawling, prostrate forb with long stems from large taproots. Plants may be several feet across and less than 6 inches high.

**Reproduction:** Seeds.

**General Description:** Leaves are hairless and triangular (deltoid), with wavy edges. All plants in the Nyctaginaceae have opposite leaf arrangement. Stems are brittle. Flowers are long, trumpet-shaped tubes 3–5 inches long, flared at the end. In this nocturnal plant, the impressive flowers open mostly late in the afternoon, stay open throughout the night, and close soon after sunrise. During the night, moths feed on the flower nectar and thereby pollinate the plant (Ajilvsgi 2003). When in full flower, this is a striking plant. Otherwise, it is visually unimpressive. Other common name: yerba de la rabia.

**Livestock and Wildlife Value:** Modestly grazed by sheep, goats, and white-tailed deer.

**Management:** A desirable forb that can be maintained by good grazing management.

**Leaf**—Leaves are long and narrow, with a prominent midrib, and are arranged oppositely along the stem.

**Summer Color**—
As you walk through a grassy pasture on a dry, hot summer day, this flower may catch your eye. Typically this plant displays its striking purple flowers all summer long. Photos by George Clendenin, USDA-NRCS.

# Narrowleaf spiderling *(Boerhavia linearifolia)*
## NYCTAGINACEAE—FOUR-O'CLOCK FAMILY

**Characteristics:** Native; perennial.

**Flower Bloom Period:** Spring–fall in response to rainfall.

**Growth Form:** Small, sparse, weak-stemmed forb from a woody taproot.

**Reproduction:** Seeds.

**General Description:** This plant is easily overlooked unless it is flowering. It is often hidden among grasses. Flowers are delicate and small, 5-lobed, and a deep, rich purple.

**Livestock and Wildlife Value:** Grazed by sheep, goats, and white-tailed deer, but provides only a small amount of forage.

**Management:** No specific management is needed for this plant.

Flower—Flowers of this plant open in the cool late afternoon and remain open until the next morning's hot sun appears. Photos by George Clendenin, USDA-NRCS.

# Scarlet musk flower *(Nyctaginia capitata)*
## NYCTAGINACEAE — FOUR-O'CLOCK FAMILY

**Characteristics:** Native; perennial.

**Flower Bloom Period:** Spring–fall in response to rainfall.

**Growth Form:** Upright multistemmed forb from a large taproot; height 12–18 inches.

**Reproduction:** Seeds.

**General Description:** A very unique and unmistakable plant. Leaves are opposite, large, triangular with wavy edges, and often sticky, hairy, and speckled with silver or gray splotches. The plant produces a bright scarlet bouquet of small tubular flowers that send out numerous attractive red filaments. The entire plant has an offensive musky odor. Other common name: devil's bouquet.

**Livestock and Wildlife Value:** Rarely, if ever, grazed by livestock or white-tailed deer.

**Management:** No specific management is needed for this plant. It is not noxious or aggressive.

Photos by George Clendenin, USDA-NRCS.

# Low menodora *(Menodora heterophylla)*
## OLEACEAE — OLIVE FAMILY

**Characteristics:** Native; perennial.

**Flower Bloom Period:** Spring–fall in response to rainfall.

**Growth Form:** Low forb that forms small colonies from a common root system. Colonies may be 1 foot across and 2–4 inches tall.

**Reproduction:** Seeds and rhizomes.

**General Description:** Blooms multiple times throughout the growing season in response to rainfall. When flowers are closed, the buds are red, but when flowers open up, the petals are bright yellow. Flowers have 5 or 6 petals. Leaves are opposite and vary from entire to deeply lobed with 3–7 lobes. Seeds are formed in twin round translucent capsules that split in half and contain 4 seeds each. Other common name: redbud.

**Livestock and Wildlife Value:** Often heavily grazed by sheep, goats, and white-tailed deer. Quail commonly eat the seeds.

**Management:** Conservative grazing management and careful attention to white-tailed deer numbers will help maintain or increase this desirable forb.

A primrose by any other name . . . The club-shaped stigma in this western primrose flower is characteristic of the genus *Calylophus*. *Oenothera* (showy evening primrose) flowers, on the other hand, always have a 4-branched (cross-shaped) stigma.

Is this a buttercup? Almost all evening primroses are called buttercups because of the cup-shaped flowers and the butter-colored pollen in the middle (e.g., showy evening primrose). In this case the whole flower is butter colored. Photos by George Clendenin, USDA-NRCS.

# Western primrose *(Calylophus hartwegii)*
## ONAGRACEAE—EVENING PRIMROSE FAMILY

**Characteristics:** Native; perennial.

**Flower Bloom Period:** Spring–early summer.

**Growth Form:** Medium to large forb from a large taproot. Size varies according to grazing intensity; in moderately grazed areas, plants will usually be less than 12 inches across and less than 6 inches tall. In lightly grazed or nongrazed areas, plants will often be 24 inches across and 12 inches tall.

**Reproduction:** Seeds and short underground rooting stems that produce offsets.

**General Description:** Grows best in shallow or rocky soils. Flowers of plants in the *Calylophus* genus are known to open in late afternoon or around sunset. Western primrose flowers sometimes stay open all day. Prior to flowering, this plant is not very impressive or noticeable; when flowering, it stands out with numerous large yellow flowers. Each flower has 4 petals, as do the flowers of all primroses. Other common name: buttercup.

**Livestock and Wildlife Value:** Western primrose is extremely palatable to sheep, goats, and white-tailed deer and is often grazed by cattle.

**Management:** Being one of the more palatable perennial forbs, it is not frequently found in moderately or heavily grazed pastures. Under very good, conservative grazing management and where white-tailed deer numbers are properly managed, this plant may be common.

Photos by
George Clendenin,
USDA-NRCS.

# Limestone gaura *(Gaura calcicola)*
## ONAGRACEAE—EVENING PRIMROSE FAMILY

**Characteristics:** Native; perennial.

**Flower Bloom Period:** Spring.

**Growth Form:** Medium to large forb from a woody taproot, sometimes forms colonies of interconnected plants. Plant is low growing prior to flowering. At flowering, long waving stalks emerge and grow to a height of 18–24 inches.

**Reproduction:** Seeds and short underground rooting stems.

**General Description:** Each flowering stalk supports numerous flowers that mature in staggered fashion from bottom to top. Flowers range from white to pink.

**Livestock and Wildlife Value:** Readily grazed by sheep, goats, and white-tailed deer, and sometimes cattle. Quail and turkeys sometimes eat the seed capsules.

**Management:** Conservative grazing management is needed to maintain or increase this plant.

Another primrose by any other name . . . The 4-branched (cross-shaped) stigma in this showy evening primrose flower is always found in flowers of the genus *Oenothera*. *Calylophus* (western primrose) flowers, on the other hand, always have a club-shaped stigma.

Leaves—Leaves are toothed to deeply pinnatifid, and alternately arranged.

A Show of Color—Showy evening primrose is commonly found in grazed pastures as well as in yards, parks, and roadsides where it may get extra moisture. It is commonly planted in wildflower gardens. Photos by George Clendenin, USDA-NRCS.

# Showy evening primrose *(Oenothera speciosa)*
## ONAGRACEAE—EVENING PRIMROSE FAMILY

**Characteristics:** Native; perennial.

**Flower Bloom Period:** Spring.

**Growth Form:** Occurs in colonies of many interconnected plants. Colonies may be 3–10 feet across and 12 inches tall when flowering.

**Reproduction:** Seeds and prolific system of rhizomes.

**General Description:** When flowering, this plant is unmistakable, with dozens of large, showy pink flowers. Each flower is composed of 4 large petals; the veins are a deep, dark pink while the remainder of the petal is soft pink. Flowers may open in the evening or morning. Moths are said to be attracted to the flowers' scent and to pollinate them at night. This plant grows best in moist, protected, and partially shaded sites, although it can also grow in other locations. Other common names: pink evening primrose, amapola, buttercup.

**Livestock and Wildlife Value:** Extremely palatable to livestock and white-tailed deer.

**Management:** Because it is so palatable to livestock and white-tailed deer, it is not frequently found in grazed pastures. Where it is present, it can be maintained by light rotational grazing.

**Other Uses:** Oil from plants in the *Oenothera* genus is an important ingredient in many cosmetics, including lipstick (Diggs, Lipscomb, and O'Kennon 1999).

Bulb or Tuber? This is a true bulb and therefore has fleshy layers or scales that store energy for the plant and surround an internal bud, just like an onion. (Cut an onion vertically in half and see!) At the bottom of the bulb is an anchor, or basal plate, that holds the bulb together and has roots emerging from it. A tuber also stores food but does not have scales, or an internal bud, like a bulb. The Oxalidaceae family has both bulbs and tubers.

Purple flowers appear in the fall and help us find Drummond's oxalis in these low, grassy areas. Our other woodsorrel (*Oxalis stricta*) has yellow flowers. Photos by George Clendenin, USDA-NRCS.

# Drummond's oxalis *(Oxalis drummondii)*
## OXALIDACEAE — WOODSORREL FAMILY

**Characteristics:** Native; perennial.

**Flower Bloom Period:** Fall.

**Growth Form:** Delicate stalks and leaves from small underground bulbs; height 4–8 inches.

**Reproduction:** Seeds and movement of bulbs by burrowing or digging animals.

**General Description:** This plant is invisible most of the year, persisting as an underground bulb until sufficient moisture is available for growth. Leaves and stems may appear soon after good spring or summer rains. The odd-looking leaves are 3-parted, with each section strongly folded in the center. Flowers are pink and have 5 petals. The leaves and stems have a strong sour taste because of the presence of oxalic acid. Other common name: woodsorrel.

**Livestock and Wildlife Value:** Plant is grazed by sheep, goats, and white-tailed deer. Bulbs are sometimes dug and eaten by turkeys, javelinas, feral hogs, and small rodents. The bulbs were one of the primary foods of Montezuma quail, which were common in the Concho Valley in the 1800s and which are still found in the Trans-Pecos.

**Management:** Plant is somewhat resistant to heavy grazing since it grows so rapidly after rain, completing its life cycle and then retreating back to an underground bulb. No specific management is needed to sustain populations of this plant.

**A Favorite among Rabbits**—Woodsorrel is a favorite delicacy for rabbits. My daughter's rabbit, "Missy," used to love to graze on this plant during her "free-range" time!

What is the difference between woodsorrel and clover? Our woodsorrel (*Oxalis*) has heart-shaped leaves, although these are not necessarily unique to *Oxalis*. True clovers (*Trifolium*) do not ever have heart-shaped leaves. Finding the true Irish shamrock is tricky. It could be *Oxalis acetosella* but is more likely *Trifolium dubium* or *T. repens* (Diggs, Lipscomb, and O'Kennon 1999). Look it up and you decide! Photos by George Clendenin, USDA-NRCS.

# Woodsorrel *(Oxalis stricta)*
## OXALIDACEAE — WOODSORREL FAMILY

**Characteristics:** Native; perennial.

**Flower Bloom Period:** Spring–fall in response to rainfall.

**Growth Form:** Low, mat-forming forb; height 3–6 inches.

**Reproduction:** Seeds, shallow rhizomes, and rooting at the nodes of stems.

**General Description:** Leaves have the general appearance of a "three-leaved clover," but the plant is unrelated to clover. Each leaflet is heart shaped. The presence of oxalic acid in the leaves gives them a strong sour flavor; hence the name "sour clover." Flowers are small and yellow, with 5 petals. Fruits are long, narrow capsules that faintly resemble miniature okra pods. More commonly found in yards, parks, and disturbed areas; less commonly found on ranches. Other common names: sheep showers, yellow woodsorrel.

**Livestock and Wildlife Value:** Grazed by sheep, goats, and white-tailed deer. Seedpods are eaten by quail, turkeys, and other birds.

**Management:** No specific management is recommended for this plant.

**Native American Uses:** Pawnees of the Missouri valley say that the buffalo were very fond of *Oxalis stricta*. They also say that the children ate both *O. stricta* and *O. violacea* (violet woodsorrel, found elsewhere in Texas) but especially preferred and ate the leaves, flowers, scapes, and bulbs of *O. violacea*. The Pawnees also pounded the bulbs and fed them to horses to make them fleet (Gilmore 1919).

**Moderation in All Things:** Woodsorrel, like spinach, contains oxalic acid, which can be dangerous to humans and animals in high concentrations. However, in small amounts, oxalis has not been known to be dangerous to humans. People have sprinkled oxalis in their salad or soup to add a tart flavoring. It is also said to be high in vitamin C and was used in the past to treat scurvy (Tull 1987).

Photos by
George Clendenin,
USDA-NRCS.

300

# White pricklypoppy *(Argemone aurantiaca)*
## PAPAVERACEAE—POPPY FAMILY

**Characteristics:** Native; annual; potentially biennial.

**Flower Bloom Period:** Spring–early summer.

**Growth Form:** Tall, upright, spiny-leaved forb from a taproot; height 18–36 inches.

**Reproduction:** Seeds.

**General Description:** Leaves are extremely spiny and an obvious blue green. Sap is bright yellow or orange. Flowers are large and attractive, with 6 delicate white petals and a large cluster of yellow stamens in the center. Seedpods are large spiny capsules that slowly split open, spilling seeds over a prolonged period.

**Livestock and Wildlife Value:** White pricklypoppy is not grazed by livestock or white-tailed deer because of the extremely spiny and possibly toxic leaves. The seeds are high in fat and protein and endure weathering very well, making them extremely valuable to quail, doves, and other birds. In fact, this is one of the top 10 plants for doves and quail in the region.

**Management:** This plant can be stimulated by soil disturbance, such as disking or the creation of fireguards. Otherwise, the plant is uncommon in pastures with a good cover of grass.

**Fruit and Leaves**—Spikes of small bright red berries form in late summer and fall. Nutritional analysis of the leaves indicates very high crude protein levels of 27–39 percent in spring. Photos by George Clendenin, USDA-NRCS.

# Pigeonberry *(Rivina humilis)*
## PHYTOLACCACEAE—POKEWEED FAMILY

**Characteristics:** Native; perennial.

**Flower Bloom Period:** Late spring–summer.

**Growth Form:** Sprawling multistemmed forb from a large central taproot; height 12–18 inches.

**Reproduction:** Seeds.

**General Description:** Pigeonberry is often seen growing in the protection of spiny bushes and under the partial shade of mesquite. It is sometimes found growing in small colonies. Flower stalks start to bloom in midsummer as compact spikes of tiny pink flowers. The leaves are ovate, sometimes with wavy margins, and are arranged alternately. All parts of the plant are considered toxic to humans (Tull 1987). Other common names: bloodberry rouge plant, rouge plant, coralito.

**Livestock and Wildlife Value:** Readily grazed by livestock and white-tailed deer. Berries are eaten by many species of birds.

**Management:** Retaining large mature mesquite and associated underbrush will help ensure proper habitat for this valuable plant. Plant is not tolerant of heavy grazing.

**Native American and Current Uses:** You can use the berries of this plant as an ink or a red dye. Do not handle with your hands, though; remember that all parts of the plant are toxic! A plant that Native Americans more likely used for dyes was the larger pokeweed, *Phytolacca americana* (Tull 1987).

A Changing Look? This photo shows a young, healthy green plant with all of its parts present. When our plantains start drying and are found later in the summer, typically all of the leaves are gone and just the flower stalk is sticking out of the ground. Photo by George Clendenin, USDA-NRCS.

# Heller's plantain *(Plantago helleri)*
## PLANTAGINACEAE—PLANTAIN FAMILY

**Characteristics:** Native; annual.

**Flower Bloom Period:** Spring.

**Growth Form:** Small forb; height 3–6 inches.

**Reproduction:** Seeds.

**General Description:** Long, narrow, hairy leaves grow in a low rosette during winter and early spring. As spring arrives, flowering stalks emerge with a cluster of small flowers. Flower petals are an unusual opaque white. Leaves have mostly parallel veins and are basal, meaning located at the base of the plant. Other common names: cedar plantain, tallow weed.

**Livestock and Wildlife Value:** This species of plantain is grazed by sheep, goats, and white-tailed deer, but not to the same extent as the more desirable red-seed plantain. Birds and small rodents eat the seeds.

**Management:** Annual forbs such as this grow in greatest abundance in bare, sparsely vegetated, or disturbed soils. Keeping a good cover of grass will reduce the abundance of these winter annuals.

**Mature plant**—This photo shows a mature, dried plant with stalks but without most of the basal leaves. Photos by George Clendenin, USDA-NRCS.

# Red-seed plantain *(Plantago rhodosperma)*
## PLANTAGINACEAE—PLANTAIN FAMILY

**Characteristics:** Native; annual.

**Flower Bloom Period:** Spring.

**Growth Form:** Small leafy forb; height 4–8 inches.

**Reproduction:** Seeds.

**General Description:** Plant forms cool-season rosettes in winter. Leaves with prominent veins are much broader than those of the other species of plantain. Leaves are covered with short, stiff hairs at maturity, giving the plant a bristly texture. Plant grows in great abundance in years with good winter and spring rains. Flowers are inconspicuous and unattractive, forming in tall spikes in late winter and early spring. Many small red seeds are produced on these stalks by late spring. These forbs are weak and shallow rooted and will desiccate and disappear quickly with the arrival of hot, dry weather. Other common name: tallow weed, Indian wheat.

**Livestock and Wildlife Value:** Prized by ranchers for the large amount of good-quality grazing it provides in some years; hence the name "tallow weed." This plant is grazed by all classes of livestock and also by white-tailed deer. However, the manna is short-lived and unreliable. Cattle are fond of grazing the mature seed stalks in summer, and thousands of the red seeds can be found in cow pies.

**Management:** A thick cover of desirable grass will reduce the probability of growing an abundance of this plant; a sparse cover of grass and bare ground will encourage production. Such winter annuals have their place in ranching, as they add green forage when most grasses are dormant; however, an overabundance probably indicates a lack of good grass cover.

Leaves—Narrow leaves are alternate and crowded toward the base of the stems. Photos by George Clendenin, USDA-NRCS.

# White milkwort *(Polygala alba)*
## POLYGALACEAE—MILKWORT FAMILY

**Characteristics:** Native; perennial.

**Flower Bloom Period:** Spring.

**Growth Form:** Small slender forb with many stems from a central taproot; height 12–18 inches.

**Reproduction:** Seeds.

**General Description:** This plant will usually go completely unnoticed until the flowers bloom on the long stalks. Flower stalk is a spike of many very small white flowers.

**Livestock and Wildlife Value:** Readily grazed by sheep, goats, and white-tailed deer in spring and summer.

**Management:** This and most other desirable perennial forbs benefit from conservative grazing management and are harmed by heavy or continuous grazing.

**Got Milk?** The Polygalaceae family gets its name from the Greek, *polys* meaning "much," and *gala* meaning "milk." In first-century Greece, it was thought that when taken as a drink, this plant would increase milk production in nursing women. Some European species were thought to increase milk production in cows (Diggs, Lipscomb, and O'Kennon 1999; Ajilvsgi 2003). The powdered root of some *Polygala* species is termed pharmaceutically "senega" or "senegrin" and used to treat a variety of respiratory ailments (Ajilvsgi 2003).

Photos by
George Clendenin,
USDA-NRCS.

# Bluets *(Hedyotis nigricans)*
## RUBIACEAE—COFFEE OR MADDER FAMILY

**Characteristics:** Native; perennial.

**Flower Bloom Period:** Spring–fall in response to rainfall.

**Growth Form:** Short multistemmed forb from a central taproot; height 4–12 inches.

**Reproduction:** Seeds.

**General Description:** Stems give rise to clusters of short leaves that taper to a point. Flowers have 4 petals and range from white to lavender to pink. Leaves are opposite, linear and threadlike, and often in clusters along the stem. Other common names: prairie bluets, fine-leaf bluets, baby's breath.

**Livestock and Wildlife Value:** Seldom grazed by livestock or white-tailed deer unless other plants are not available.

**Management:** This plant requires no special management.

**Fruits and Leaves**—Fruits are strangely shaped capsules that faintly resemble a pair of pants (breeches) with the legs pointing upward. Leaves are linear, with slightly toothed margins, and alternately arranged. The leaves are strongly dotted with glands, which are visible when held up to a bright light. Photos by George Clendenin, USDA-NRCS.

# Dutchman's breeches *(Thamnosma texana)*
## RUTACEAE—CITRUS FAMILY

**Characteristics:** Native; perennial.

**Flower Bloom Period:** Spring.

**Growth Form:** Sprawling multistemmed forb from a large central taproot; height 6–12 inches.

**Reproduction:** Seeds.

**General Description:** This plant is more often smelled than seen. The entire plant has a strong citrus odor when the leaves are bruised; the smell can be noticed when vehicles drive over it, or humans or livestock walk on it. Flowers are small and yellow and remain closed.

**Livestock and Wildlife Value:** Sheep and goats moderately like the taste of this plant, even though it can be toxic and cause sensitivity to sunlight. Most poisoning cases have occurred when sheep or goats were placed in pastures that for years had been grazed only by cattle. White-tailed deer also eat this plant occasionally. Quail and other songbirds eat the seeds.

**Management:** Even though the plant is listed as toxic, it causes no serious or widespread problems in this region.

**Toxic Agent:** Psoralens (strong photosensitizing agents) (Hart et al. 2010).

**Toxic Description:** Sheep consuming about 1 percent of their body weight of this plant have shown signs of primary photosensitization within 24–48 hours. Cattle and goats also have shown primary photosensitization when grazing areas with large amounts of this plant. Humans who get sap from this plant on their hands may also get severe sunburn on those areas. See *Toxic Plants of Texas* (Hart et al. 2010) for livestock clinical signs and more detailed information.

Flower—White and showy, sometimes with light purple veins extending from the throat. Photos by George Clendenin, USDA-NRCS.

# Penstemon *(Penstemon cobaea)*
## SCROPHULARIACEAE—FIGWORT FAMILY

**Characteristics:** Native; perennial.

**Flower Bloom Period:** Spring.

**Growth Form:** Forms clump of upright multistemmed wildflowers; height to 2 feet.

**Reproduction:** Seeds.

**General Description:** One of our more beautiful wildflowers in the Concho Valley. Plant is covered with fine hairs. Leaves are long, tapered, and lance-like and are finely toothed along the margin. Unlike some of our other drought-hardy flowering perennials, penstemon has a delicate nature and prefers some good spring rains before making an appearance. Other common names: beardtongue, wild foxglove, foxglove, cobaea penstemon.

**Livestock and Wildlife Value:** Extremely palatable to livestock and white-tailed deer.

**Management:** Because it is so palatable to livestock and white-tailed deer, it is not frequently found in grazed pastures. It is more frequent along roadsides in calcareous soils.

Photos by George Clendenin, USDA-NRCS.

# Jimsonweed *(Datura stramonium)*
## SOLANACEAE—NIGHTSHADE FAMILY

**Characteristics:** Native; annual.

**Flower Bloom Period:** Spring–fall in response to rainfall.

**Growth Form:** Sprawling bushy annual; height to 5 feet.

**Reproduction:** Seeds.

**General Description:** A rank-smelling annual, native throughout North America. This is one of the more dangerously toxic plants of the Concho Valley. Large, showy flowers are trumpet shaped (funnelform), 3½–5 inches long, and white to lavender. Mature leaves are alternate, toothed to shallowly lobed, and very large. Fruits are capsules about 2 inches long, fleshy at first but becoming covered with stiff, odd-looking prickles. Seeds are numerous, flattened, dark brown to black, and minutely pitted. I have never seen more historic writings or read about uses by more peoples than for this single dangerous plant. Other common names: Indian apple, thorn apple, pricklyburr, Jamestown weed.

**Livestock and Wildlife Value:** Extremely toxic to all classes of livestock and humans. See toxicity information.

**Management:** Grubbing or herbicides easily control this plant. Exercise caution when treating jimsonweed with 2,4-D, as it makes the plant more palatable. Check watering areas and pens carefully for this plant before stocking with livestock.

**Historical Note:** The name "jimsonweed" is an alteration of the name "Jamestown weed," which arose because of an incident outside colonial Jamestown in 1676. British soldiers who were sent to Jamestown in 1676 to quell Bacon's rebellion gathered the young sprouts of *D. stramonium* and ate them as a "pottage." They suffered the ill effects of this narcotic poison for 11 days before regaining their composure, without memory of the time passed (Stevenson 1908).

**Toxic Agent:** Tropane alkaloids (atropine, scopolamine, hyoscyamine) (Hart et al. 2010).

**Toxic Description:** Poisons all livestock and people. All parts of the plant are toxic, including the seeds. Consumption of 0.1 percent of the body weight, or as little as 10–14 ounces of the plant, can kill cattle. See *Toxic Plants of Texas* (Hart et al. 2010) for livestock clinical signs and more detailed information.

Photos by
George Clendenin,
USDA-NRCS.

# Ground cherry *(Physalis cinerascens)*
## SOLANACEAE—NIGHTSHADE FAMILY

**Characteristics:** Native; perennial.

**Flower Bloom Period:** Primarily spring.

**Growth Form:** Sprawling forb; height 6–12 inches.

**Reproduction:** Seeds.

**General Description:** Ground cherry stems are weak and usually fall over from their own weight. The broad leaves have wavy edges and are covered with dense stellate (star-shaped under magnification) hairs. Flowers are pale yellow, sometimes with brown spots at the center. Flowers and fruits hang upside down. A single berry with several seeds is formed inside a leafy husk. Other common name: prairie ground cherry.

**Livestock and Wildlife Value:** Grazed by sheep, goats, and white-tailed deer. Turkeys, quail, and other birds eat the berries.

**Management:** Conservative grazing management favors this plant.

**Native American Uses:** The root of this plant was used medicinally in smoke treatment. A dressing for wounds was also made from it (Gilmore 1919).

Photos by
George Clendenin,
USDA-NRCS.

# Western horsenettle (*Solanum dimidiatum*)
## SOLANACEAE—NIGHTSHADE FAMILY

**Characteristics:** Native; perennial.

**Flower Bloom Period:** Spring–summer.

**Growth Form:** Coarse, leafy forb from a large taproot; height 12–18 inches.

**Reproduction:** Seeds.

**General Description:** Stem is covered by numerous sharp spines. The large leaves have lobed edges and are covered in hairs. Spines are also present on the lower veins of the leaves. Flowers can be white, blue, or purple. Hard berries are yellow at maturity and up to 1 inch in diameter. Other common name: potato-weed, tread salve.

**Livestock and Wildlife Value:** Western horsenettle is very unpalatable and is consumed only when other forage is unavailable. Leaves and berries are toxic to livestock, especially cattle. Turkeys frequently eat the fruits with no ill effects.

**Management:** Rotate cattle from pastures infested with western horsenettle to help prevent "Crazy Cow Syndrome." Take care to prevent baling fruiting plants into hay. The roots are very deep and strong and not easily killed by plowing. This plant can be controlled by selective spot application of herbicides.

**Toxic Agent:** Glycoalkaloids and calystegines, in varying amounts, in all parts of the plant. The highest concentrations are in the ripe fruits (Hart et al. 2010).

**Toxic Description:** Affects cattle, sheep, goats, and horses. Low-level intake of calystegines over several months is probably responsible for a nervous condition called "Crazy Cow Syndrome," which, based on current information, is limited to two geographical areas: Real County and parts of surrounding counties, and a larger area encompassing Glasscock, Menard, and Taylor Counties. See *Toxic Plants of Texas* (Hart et al. 2010) for livestock clinical signs and more detailed information.

Photos by George Clendenin, USDA-NRCS.

# Silverleaf nightshade *(Solanum elaeagnifolium)*
## SOLANACEAE — NIGHTSHADE FAMILY

**Characteristics:** Native; perennial.

**Flower Bloom Period:** Spring–summer.

**Growth Form:** Coarse, leafy forb from a woody taproot; height 12–18 inches.

**Reproduction:** Seeds and creeping rootstalks.

**General Description:** Stems are covered by sharp spines. Leaves are long and narrow, grayish green on the upper surface and silver on the lower surface. Flowers are purple with large yellow stamens in the center. Hard berries are green initially and turn yellow at maturity. Berries are about ½ inch in diameter.

**Livestock and Wildlife Value:** Silverleaf nightshade is very unpalatable and is consumed only when other forage is unavailable. Toxic to livestock; see below. Turkeys frequently eat the fruits with no ill effects.

**Management:** Care should be taken, as this plant may be baled with hay. Do not feed from the ground where many ripe nightshade fruits are available. This plant can be controlled by selective spot application of herbicides. The roots are very deep and strong and not easily killed by plowing. In fact, mechanical ground disturbance may increase this plant.

**Native American Uses:** The Zuni word for this plant is *Ha'watapa* (*ha* or *ha'li*, "leaf"; *wa'tapa*, "prickly"). The chewed root was placed in the cavity of an aching tooth. It is also said that the Zuni regarded curdled goat's milk together with the berries as a delicious beverage (Stevenson 1908). (WARNING: This is a toxic plant, including the berries; do not consume.)

**Toxic Agent:** The glycoalkaloid solanine (Hart et al. 2010).

**Toxic Description:** This plant has reportedly poisoned horses, sheep, cattle, goats, and humans. However, sheep and goats are more resistant than cattle. In some controlled experiments, goats were not poisoned at all. The leaves and fruits are toxic at all stages of maturity, the highest toxicity being in the ripe fruits. Animals can be poisoned by eating 0.1 to 0.3 percent of their weight in silverleaf nightshade. See *Toxic Plants of Texas* (Hart et al. 2010) for livestock clinical signs and more detailed information.

**Why Name It after a Buffalo?**—Presumably, buffalobur got its common name when roaming buffalo would get the spiny fruits caught in their hair while roaming and foraging the prairies (Diggs, Lipscomb, and O'Kennon 1999; Ajilvsgi 2003). Photos by George Clendenin, USDA-NRCS.

# Buffalobur *(Solanum rostratum)*
## SOLANACEAE—NIGHTSHADE FAMILY

**Characteristics:** Native; annual.

**Flower Bloom Period:** Spring–fall in response to rainfall.

**Growth Form:** Low, bushy forb; height 12–18 inches.

**Reproduction:** Seeds.

**General Description:** Buffalobur is a warm-season forb. Leaves are broad and spiny, with many deep lobes. Stems are extremely spiny. Flowers are yellow, with pleated petals. Seedpods are covered with spines. Plant grows in disturbed areas and is a common weed in cropland. Since it is an annual, its abundance will vary considerably from year to year.

**Livestock and Wildlife Value:** Buffalobur is toxic to livestock, although it is not often consumed because of its unpalatable nature and probable mechanical injury to the mouth from spines. The seeds are frequently eaten by doves and quail and sometimes make up the majority of their diet during fall. There is no known forage value to white-tailed deer.

**Management:** This plant will be uncommon in good grass cover but may be abundant in overgrazed or recently disturbed areas and following mechanical brush control. Landowners interested in increasing the abundance of food for quail and doves sometimes disk areas to encourage the production of this and other seed-bearing plants.

**Native American Uses:** The Zuni word for this plant is *Mo´kĭachipa* (*mo* or *m´oli,* "round" or "roundish," denoting the pod; *kĭa´chipa,* "prickly"). "A pinch of the powdered root is put into a small quantity of water and the infusion drunk to relieve sick stomach. It does not act as an emetic" (Stevenson 1908).

**Toxic Agent:** The glycoalkaloid solanine (Hart et al. 2010).

**Toxic Description:** This plant has reportedly poisoned horses, sheep, cattle, and goats. However, sheep and goats are more resistant than cattle. In some controlled experiments, goats were not poisoned at all. The leaves and fruits are toxic at all stages of maturity. Animals can be poisoned by eating 0.1 to 0.3 percent of their weight in buffalobur. See *Toxic Plants of Texas* (Hart et al. 2010) for livestock signs and more detailed information.

The small flowers usually hang downward and have 5 white petals.

"Triquetrum" means 3-cornered, like this leaf; however, leaf shapes are extremely variable in shape and size, and often many different-shaped leaves are present on the same plant. Photos by George Clendenin, USDA-NRCS.

# Texas nightshade *(Solanum triquetrum)*
## SOLANACEAE—NIGHTSHADE FAMILY

**Characteristics:** Native; perennial.

**Flower Bloom Period:** Spring–fall in response to rainfall.

**Growth Form:** Viny multistemmed forb (or woody vine) from a large woody base. Height can vary greatly depending on setting. When supported by bushes or fences, it grows to a height of over 6 feet. Without such support, the height is less than 12 inches.

**Reproduction:** Seeds and underground rooting stems that form numerous root sprouts.

**General Description:** Usually seen growing in the protection of spiny bushes or brush piles, which are used for support. The long, trailing stems are sometimes nearly naked of leaves, appearing as a mass of green stems. Berries are bright red and ¼ inch in diameter.

**Livestock and Wildlife Value:** Texas nightshade is extremely palatable to all classes of livestock and to white-tailed deer. Turkeys, quail, and many other birds are fond of the berries. This nightshade is the exception to the general rule that nightshades are poisonous, undesirable weeds.

**Management:** This plant can be maintained by careful attention to grazing management and control of white-tailed deer numbers.

The Gingerbread Man? Some say the 5-petalled flower resembles a gingerbread man. Can you see one? Notice that each petal has a notch in it.

Community Gathering—The verbena pictured here is also intimately sharing its space with one of our ragweeds, also with finely dissected leaves. Can you tell which one? Photos by George Clendenin, USDA-NRCS.

# Prairie verbena *(Glandularia bipinnatifida)*
(Sy = *Verbena bipinnatifida*)
## VERBENACEAE — VERVAIN OR VERBENA FAMILY

**Characteristics:** Native; perennial.

**Flower Bloom Period:** Primarily spring–early summer.

**Growth Form:** Large, sprawling, much-branched forb; diameter 3 feet, height 1 foot.

**Reproduction:** Seeds.

**General Description:** Stems trail across the ground. Leaves are much dissected. Leaves, stems, and flower stalks are covered by hair. The flower heads are large and showy compared to those of other species of verbena. This is one of the most recognizable and popular wildflowers in the region. Found throughout Texas but mainly in the western two-thirds of the state. Other common names: Dakota verbena, Dakota vervain, prairie vervain.

**Livestock and Wildlife Value:** Prairie verbena is occasionally grazed by livestock and white-tailed deer but is not considered a desirable forage plant. Seeds are occasionally eaten by birds.

**Management:** Most commonly found in disturbed areas. Often abundant for a few years following mechanical brush control, but declines as grass cover improves. A chronic abundance of this plant year after year probably indicates an overgrazing problem.

**Pink Makes Yellow?** Prairie verbena is known to make an outstanding bright yellow dye. The whole plant, with flowers, is used in making the dye (Tull 1987).

**"Leaf" It to the Experts:** The leaves of prairie verbena are variable in shape but are mostly much dissected. Many leaves are bipinnatifid, giving the plant its species name. The leaves are considered bipinnatifid, rather than bipinnately compound, because they are simple, meaning they really consist of just one part or piece rather than many parts.

Photos by George Clendenin, USDA-NRCS.

# Pink verbena *(Glandularia pumila)*
(Sy = *Verbena pumila*)
## VERBENACEAE—VERVAIN OR VERBENA FAMILY

**Characteristics:** Native; annual.

**Flower Bloom Period:** Spring.

**Growth Form:** Low, sprawling forb; height 6–12 inches.

**Reproduction:** Seeds.

**General Description:** Pink verbena, an annual, is sometimes confused with prairie verbena, a perennial. The differences in the leaves will help to tell them apart. Unlike prairie verbena (*G. bipinnatifida*), pink verbena (*G. pumila*) leaves are 3-parted. They are both much dissected, but those of pink verbena are more finely dissected. Pink verbena has the same 5-petaled pink to lavender flowers, but the corollas (collective term for the petals) are smaller than those of prairie verbena. The leaves, stems, and flower stalks are covered by hair. The stems generally trail across the ground. Found throughout Texas but mainly in the western two-thirds of the state. Other common name: pink vervain.

**Livestock and Wildlife Value:** Pink verbena is occasionally grazed by livestock and white-tailed deer but is not considered a desirable forage plant. Seeds are occasionally eaten by birds.

**Management:** No specific management is recommended for this plant.

Photos by George Clendenin, USDA-NRCS.

# Texas frogfruit *(Phyla nodiflora)*
## VERBENACEAE — VERVAIN OR VERBENA FAMILY

**Characteristics:** Native; perennial.

**Flower Bloom Period:** Spring–fall in response to rainfall.

**Growth Form:** Low-growing, stemmy plant that forms large colonies; height 3–6 inches.

**Reproduction:** Seeds and runners that root at each node.

**General Description:** Easily recognized by the long, branching stems that trail across the ground, rooting at each node and forming a tangled mat. Tiny flowers form in rings on the flower head. Leaves have saw-toothed serrations and are opposite, as is characteristic in the Verbenaceae family. Found in wet areas, where it helps protect soil and control erosion. Also found in dry creek beds of the Concho Valley. Other common names: frogfruit, hierba de la Virgen María, turkey tangle.

**Livestock and Wildlife Value:** Occasionally grazed by livestock and white-tailed deer, and frequently visited by butterflies.

**Management:** No specific management practices are needed.

Photos by
George Clendenin,
USDA-NRCS.

# Slender vervain *(Verbena halei)*
## VERBENACEAE — VERVAIN OR VERBENA FAMILY

**Characteristics:** Native; perennial.

**Flower Bloom Period:** Primarily spring.

**Growth Form:** Slender, upright plant; height 6–12 inches.

**Reproduction:** Seeds.

**General Description:** Long flowering spikes with small blue-lavender flowers up to ¼ inch across. Rigid, hairy stems, sometimes 1 to several, then mostly 4-angled. Leaves are sparse, varying from deeply lobed in lower portion of plant to linear or oblanceolate in upper portion of plant. Other common name: Texas vervain.

**Livestock and Wildlife Value:** Occasionally grazed by livestock and white-tailed deer but not considered a desirable forage plant. Seeds are occasionally eaten by birds.

**Management:** No specific management is recommended for this species.

**A Charmed Plant:** Although slender vervain may be slight and rather unimpressive compared to other verbenas, it has a rich history of being regarded as a powerful charm against witches. (Ironic: a charm *against* witches?) It was also used as a medicinal herb in early times (Ajilvsgi 2003).

Flowers—Small, yellow, and composed of 5 petals, flowers eventually develop into the spiny fruit. Fruit—More appropriately called a bur, the fruit resembles the head of a 2-horned goat; thus the common name. The fruit actually starts out circular and spiny, with 5 sections that break apart at maturity, forming several of the wicked burs. Each section has 2–4 seeds. Photo by George Clendenin, USDA-NRCS.

# Goathead *(Tribulus terrestris)*
## ZYGOPHYLLACEAE — CALTROP FAMILY

**Characteristics:** Introduced; annual.

**Flower Bloom Period:** Spring–fall in response to rainfall.

**Growth Form:** Low, trailing weed that forms large mats; height only a few inches.

**Reproduction:** Seeds are formed inside burs, which become attached to the feet of passing humans or animals, thereby aiding the distribution of the seeds.

**General Description:** This is a very familiar weed to anyone who has ever tried to walk barefoot through a patch of it. It is also the main reason country kids always have flat tires on their bikes! Many long stems radiate out from a central taproot. The opposite leaves are hairy and divided into 4 to 8 pairs of leaflets, each pair slightly smaller than the pair before it (more or less). This undesirable plant is a native of Europe and is a perfect example of the potential danger of exotic plant introductions. Other common names: puncturevine, abrojo de flor amarillo.

**Livestock and Wildlife Value:** Toxic to livestock and presumably unpalatable.

**Management:** Goathead grows in disturbed areas but does not grow in areas with good grass cover. It is seldom seen on ranches but is common in "ranchettes," school yards, vacant lots, and parks. Recommended methods of control include herbicides and improvement in grass cover. Caution should be taken if treating this plant with 2,4-D, as this chemical increases nitrate accumulation in the plant.

**Toxic Agent:** Causes hepatogenous photosensitization in sheep and possibly cattle.

**Toxic Description:** All parts of the plant are toxic at all growth stages, but wilted plants are the most toxic. The plant is also mechanically dangerous because the burs can create lesions in the mouth or feet. Goathead can also accumulate high levels of nitrate. See *Toxic Plants of Texas* (Hart et al. 2010) for livestock clinical signs and more detailed information.

Fall Color—Flameleaf sumac gets its common name from the reddish color of its fall foliage. This is common for several species of sumac. This photo also shows the flameleaf plant growing out of an algerita bush, which probably afforded protection for the flameleaf seedling for the first several years.

    A Reason for the Season? Many *Rhus* species have early "fall color," in which their leaves turn dark to brilliant red. Flameleaf is the only Concho Valley sumac that turns color. This has been known to serve as a "foliar fruit flag," thereby attracting birds to the tree's fruit (Diggs, Lipscomb, and O'Kennon 1999).

Pinnately Compound Leaves—These have 5 or more leaflets. The main stem (rachis) of this unique compound leaf has a leafy wing that makes it appear almost flattened.

Summer Color—Green foliage on a spindly tree may make this specimen easy to miss. Photos by George Clendenin, USDA-NRCS.

## SHRUBS AND TREES

# Flameleaf sumac *(Rhus copallinum)*
### ANACARDIACEAE — SUMAC FAMILY

**Characteristics:** Native; deciduous.

**Growth Form:** Large deciduous shrub, often grows in colonies; normal height 8–12 feet, but sometimes grows into small tree up to 20 feet tall.

**Reproduction:** Seeds and root sprouts.

**General Description:** Flameleaf sumac is fairly common in the central Edwards Plateau but less common in our western part of the plateau. Twigs are reddish and covered by very fine fuzz. Leaves turn bright orange and red in fall. Many tiny white flowers occur in large clusters at the tips of branches. After pollination, seedheads begin to form, maturing in the fall. The mature seedhead is composed of hundreds of dry red fruits. Fruits remain intact for several months. This plant is a prolific root sprouter, especially following fire. Fire also stimulates germination of the seeds.

**Livestock and Wildlife Value:** Readily browsed by goats and white-tailed deer. Fruits are eaten by many species of birds.

**Management:** Flameleaf sumac may not be found in pastures where goat or white-tailed deer numbers are high. Proper management of browsing animals is key to maintaining this plant. Fire can be used to encourage and stimulate it. Care should be taken to protect this plant when planning mechanical or chemical brush control.

**Flowers**—Clusters of tiny white flowers appear in late winter, often before the leaves appear.

**Pinnately Compound Leaves**—These have 5–9 very small leaflets. Notice the same winged stem (rachis) as we see with flameleaf sumac.

**Community**—Littleleaf sumac is deciduous, and the leaves do not turn red in the fall, as with some other species of *Rhus*. Photos by George Clendenin, USDA-NRCS.

# Littleleaf sumac *(Rhus microphylla)*
## ANACARDIACEAE — SUMAC FAMILY

**Characteristics:** Native; deciduous.

**Growth Form:** Large, intricately branched, deciduous shrub; height 10 feet, diameter to 15 feet.

**Reproduction:** Seeds.

**General Description:** Littleleaf sumac is fairly common in some areas and may be the dominant shrub. Grows best in moderately deep soil; less often found in very deep or very shallow soil. Clusters of yellow, orange, or red fruits appear in spring. Other common name: desert sumac.

**Livestock and Wildlife Value:** Browsed by goats and white-tailed deer but not highly preferred. Because of its abundance, it may be the primary browse plant in some areas. Fruits are eaten by many species of birds as well as goats, white-tailed deer, and small mammals. Bushes provide very good protective cover for small songbirds and excellent loafing cover for quail.

**Management:** This plant should be protected during brush-control operations if wildlife is an important consideration. Littleleaf sumac is very vulnerable to damage and death from the use of some common brush-control herbicides. Like most other shrubs, it resprouts vigorously after fire. Prescribed burning can be used to increase the nutritional value of browse on this and other shrubs.

**Leaves**—Palmately compound with 3 leaflets. Plants usually flower before leaves appear.

**Community**—Skunkbush sumac often grows in clumps or mottes under the canopy of other trees, especially live oak. Photos by George Clendenin, USDA-NRCS.

# Skunkbush sumac *(Rhus trilobata)*
(Sy = *Rhus aromatica* var. *flabelliformis*)
**ANACARDIACEAE — SUMAC FAMILY**

**Characteristics:** Native; deciduous.

**Growth Form:** Medium-sized deciduous shrub; diameter to 6 feet, height to 4 feet.

**Reproduction:** Seeds. Skunkbush can be propagated by seeds and cuttings (Nokes 2001).

**General Description:** Small whitish-yellow flowers emerge in clusters in early spring, followed by fruit clusters in early summer. Fruits are dry, with a bright red husk covered in fine fuzz. Usually grows in shallow soils, rarely in deep soil.

**Livestock and Wildlife Value:** Skunkbush sumac is browsed by goats and white-tailed deer. Fruits are eaten by many species of birds as well as goats, white-tailed deer, and small mammals. Bushes provide good protective cover for small songbirds and loafing cover for quail.

**Management:** Like most other desirable shrubs, this plant is maintained by paying attention to the number of browsing animals. This shrub should normally be protected when conducting mechanical brush control. Plant is susceptible to some herbicides commonly used for brush control.

Community—Clusters of tiny white flowers appear in summer followed by fuzzy dry red fruits that mature in fall. The combination of bright glossy green leaves and bright red fruits is striking and attractive. Photos by George Clendenin, USDA-NRCS.

# Evergreen sumac *(Rhus virens)*
*(Sy = R. sempervirens)*
## ANACARDIACEAE—SUMAC FAMILY

**Characteristics:** Native; evergreen.

**Growth Form:** Medium to large evergreen shrub; height 6–8 feet, sometimes grows to size of small tree.

**Reproduction:** Seeds.

**General Description:** A true evergreen shrub; glossy leaves are composed of several oval leaflets, with reddish twigs that are covered by fine fuzz. Evergreen sumac grows in canyons and headers and on steep hillsides, almost always in rocky soils.

**Livestock and Wildlife Value:** Evergreen sumac is browsed by goats and white-tailed deer but is not highly preferred. Fruits are eaten by many species of birds as well as goats, white-tailed deer, and small mammals. Bushes provide protective cover for small songbirds.

**Management:** Proper control of goat and white-tailed deer numbers is all that is needed to maintain this desirable shrub. Unintentional damage by brush control is seldom a problem because of the rough terrain where this plant is found.

**How to Grow:** Evergreen sumac is a beautiful native evergreen shrub that can make an excellent addition to a landscape or garden setting. It can be propagated by seeds fairly successfully. It has lower success as a cutting, as deciduous sumacs root more easily than evergreen sumacs (Nokes 2001). See Nokes (2001) for detailed instructions on how to grow. I have also seen this shrub for sale in select native nurseries.

Community—Willow baccharis is often confused with salt cedar. Unlike salt cedar, willow baccharis is a native plant, and a natural part of the rangeland landscape. Photos by George Clendenin, USDA-NRCS.

# Willow baccharis *(Baccharis neglecta)*
## ASTERACEAE—SUNFLOWER OR DAISY FAMILY

**Characteristics:** Native; partially evergreen.

**Growth Form:** Upright, multistemmed brushy shrub; height 6–12 feet.

**Reproduction:** Seeds are attached to fine bristles that blow in the wind.

**General Description:** This shrub is completely unrelated to willow and has a rather plain appearance, but when it flowers some consider it attractive. Leaves are small and narrow. Large plumes of small white flowers appear in fall on separate male and female plants. Seeds are produced in profusion and are dispersed by wind. Plant usually grows in moist or seasonally wet areas. Other common names: dryland willow; Roosevelt weed.

**Livestock and Wildlife Value:** Willow baccharis is seldom browsed by goats or white-tailed deer except in extreme situations. Butterflies of many species use the flowers in fall, and this plant is considered an important food plant for migrating monarch butterflies.

**Management:** This shrub is seldom abundant in dry upland pastures but can become thick along creeks. The temptation is often to control the plant since it has minimal food value to livestock or white-tailed deer. However, it should normally be left in place along creeks, where it helps protect banks and resist erosion during flooding. The upright trunks help slow the velocity of floodwater and trap sediment.

Edible Leaves—I have personally enjoyed selecting and eating these fresh, tender, reddish leaves with my family on nature walks through our back pasture. They are delicious by themselves and also make a delightful addition to a fresh garden salad. The freshest edible leaves are red, not green, and do not yet have sharp edges. These edible leaves are usually available after nice spring (and sometimes fall) rains. Be careful eating large quantities; too much may be harmful, as the sour taste comes from oxalic acid.

Fruit—The berries of *Berberis trifoliolata*, as well as those of *B. swaseyi* and *B. haematocarpa,* have all been used to make jellies, pies, cobblers, and a beverage (Tull 1987).

For the Birds—Painted buntings love the fruit of algerita plants and seek out areas where these plants are present. Photos by George Clendenin, USDA-NRCS.

# Algerita *(Berberis trifoliolata)*
## BERBERIDACEAE — HOLLY FAMILY

**Characteristics:** Native; evergreen.

**Growth Form:** Medium-sized, evergreen spiny shrub; height 4–8 feet and equal or greater diameter.

**Reproduction:** Seeds.

**General Description:** Evergreen leaves are composed of 3 holly-like leaflets, each with 5 very sharp spines. Mature rigid leaves are grayish green, while young new leaves are soft and reddish. Wood is bright yellow. Yellow flowers start to bloom during late winter. Their fragrance is very sweet and almost overpowering when many plants are in bloom. If the flowers are not damaged by a late freeze, bright red berries mature by late spring. Other common name: agarita.

**Livestock and Wildlife Value:** The new, soft, tender leaves are browsed for a short time by goats, sheep, and white-tailed deer. As the leaves mature and the spines harden, animals stop browsing. The flower buds are eaten by quail, and the red berries are eaten by many kinds of songbirds and turkeys. The spiny branches offer protection for grasses, and turkeys commonly nest at the perimeter of algerita.

**Management:** This shrub is seldom heavily browsed except in the most extreme cases of very high goat numbers. If this plant is hedged by browsers, it is a sign of too many animals and indicates the need to reduce goat and white-tailed deer numbers. This shrub sometimes becomes too abundant in the eyes of some landowners. In this case the best management is prescribed burning, which will often burn the plants to the ground and greatly reduce the canopy. The bushes will resprout following fire but will not return to their original size for many years. Algerita has a very important ecological role as a protective nursery plant. It is extremely common to find other desirable shrubs or trees growing up out of an old algerita bush. Also at the edge of these bushes it is very common to find remnants of the best grasses and forbs growing where they are afforded some protection from heavy grazing.

Fruit—Clusters of bright red berries mature in fall and may stay attached through winter.

Leaves—Leaves are grayish green on the top and silvery green on the bottom. Because of this unique coloration, white honeysuckle can be noticed from a great distance. Photos by George Clendenin, USDA-NRCS.

# White honeysuckle *(Lonicera albiflora)*
## CAPRIFOLIACEAE—HONEYSUCKLE FAMILY

**Characteristics:** Native; deciduous.

**Growth Form:** Medium to large evergreen shrubby vine, sometimes with twining branches; height low or up to 10 feet.

**Reproduction:** Seeds.

**General Description:** Leaves occur in opposite matching pairs along the stem. Current year's stems may be 2–3 feet in length. Older stems have shaggy gray peeling bark. Clusters of showy white cream-colored flowers form in early spring. Japanese honeysuckle (*L. japonica*) is an introduced cultivated species that has escaped along the Concho River. The blooms of Japanese honeysuckle are very ornate compared to those of white honeysuckle. Other common name: bush honeysuckle.

**Livestock and Wildlife Value:** Preferred and desirable browse for livestock and white-tailed deer. The berries are eaten by many species of birds. A large tangled shrub provides good cover for small birds. The berries are considered toxic for human consumption.

**Management:** This and other preferred browse plants can be maintained by careful attention to livestock and deer management. Low animal numbers and rotational grazing provide periodic rest that helps promote the health and productivity of this plant. It is often found growing out of other shrubs, which afford it protection from excess browsing.

Fruit—Plant becomes more recognizable by its odd fruits, which have 4 leafy wings that turn straw colored at maturity.

Leaves—Leaves are long and narrow, with several sometimes growing from the same location. Photos by George Clendenin, USDA-NRCS.

# Fourwing saltbush *(Atriplex canescens)*
## CHENOPODIACEAE — GOOSEFOOT FAMILY

**Characteristics:** Native; evergreen.

**Growth Form:** Leafy evergreen shrub; height 4–6 feet and of equal or greater diameter.

**Reproduction:** Seeds.

**General Description:** Leaves are grayish green, giving the entire plant a distinctive color, even from a distance. Male plants begin to produce pollen in summer. Female plants begin to form fruits in late summer or fall. Fruits have 4 leafy wings that turn straw colored at maturity. Plant grows in deep soil, including saline soils in the western and northern part of the Concho Valley. Other common name: chamise.

**Livestock and Wildlife Value:** Very good browse plant, frequently eaten by goats, sheep, cattle, and white-tailed deer. Since the leaves are evergreen, the availability of winter browse is especially valuable. The plant provides cover for birds and screening for a wide variety of wildlife.

**Management:** Fourwing saltbush is best maintained by good grazing management. If heavy hedging is noted, reductions in animal numbers may be needed. Seeds are commercially available for reseeding projects. Germination and successful establishment are variable.

**Native American Uses:** The Zuni word for this plant is *Ke´mawe* (*ke,* "weed"; *ma´we,* "salt"). "The dried root and blossoms are ground separately and the two powders combined. Moistened with saliva, this mixture is employed externally to cure ant bites. When the powder is not at hand the fresh blossoms, bruised, are applied." Fourwing saltbush was also used ceremonially to bless the rabbit hunt (Stevenson 1908).

Fruit—The blue "fruit" (botanically, female cones) of an Ashe juniper tree, also indicating a female plant.

Bark—The typical fibrous bark of a juniper tree.

Growth Form—Ashe juniper typically grows straighter and looks more "treelike" than its cousin redberry juniper. This allows for its popular use in the Hill Country for cedar stays and posts used in fence construction. This tree growing by itself is uncommon, as most juniper thickets are dense, frequently with no grass in the understory. Photos by George Clendenin, USDA-NRCS.

# Ashe juniper *(Juniperus ashei)*
## CUPRESSACEAE—CYPRESS FAMILY

**Characteristics:** Native; evergreen.

**Growth Form:** Dense evergreen shrub or small tree; mature height 15–25 feet. Within the Concho Valley area, Ashe juniper usually exists on steep slopes and in deep canyons and headers. It is much less common than redberry juniper.

**Reproduction:** Seeds.

**General Description:** Ashe juniper (*J. ashei*) and redberry juniper (*J. pinchotii*) may look similar but are extremely different in terms of growth characteristics and management. Therefore, it is important to be able to tell the difference between the two. If berries are present, the difference is obvious (Ashe juniper has blue berries), but only for the female plants. Without berries, identification can sometimes be challenging. Ashe juniper foliage is a deeper green, while that of redberry juniper is a lighter green. Ashe juniper leaves lack the faint white resin deposits that are present on the leaves of redberry juniper. Ashe juniper also has a much denser shading canopy than does redberry juniper; this has important implications when it comes to grass growth beneath juniper. Other common names: blueberry cedar, mountain cedar, cedar.

**Livestock and Wildlife Value:** This plant is often underrated for its value to livestock and wildlife. Goats, white-tailed deer, and sheep browse the leaves, especially in winter when other green vegetation is scarce. The browse is not preferred over that of many other shrubs, but it is widely available in large quantity. During a dry winter, juniper often makes up a large part of the diet of goats and white-tailed deer, even on well-managed rangeland. The berries are extremely valuable to many species of birds, but this is a mixed blessing since birds are also responsible for spreading juniper seeds in their droppings. Ashe juniper in moderation also provides important cover for wildlife and a windbreak for livestock.

Shearing (above) and prescribed burning (below) are two of the most common methods for managing excessive Ashe juniper growth. Be careful, though! This form of management works only because Ashe juniper does not resprout after effective top removal. This method does *not* work for the resprouting redberry juniper (*Juniperus pinchotii*). Photos by Dee Ann Littlefield, USDA-NRCS.

**Management:** The abundance of juniper has increased markedly in the region over the last 50 to 100 years. Control of excess juniper has been a never-ending goal of many ranchers for several generations.

When the density of juniper becomes excessive, it interferes with livestock production. Dense Ashe juniper reduces forage production and plant diversity and makes livestock handling and management very difficult. Ashe juniper is considered one of the four main brush problems in the Concho Valley (along with redberry juniper, mesquite, and prickly pear). Control of Ashe juniper is easy, compared to that of other species, since it does not resprout from the stump. In fact, Ashe juniper is the only shrub or tree in the region that does not resprout from the stump. Trees are frequently cut near ground level with a chain saw, hydraulic shears, or other method. Grubbing with larger equipment is also used. The best way to control small plants and reduce future infestation is with fire, which is especially effective in killing small plants up to a height of 3 feet. Small plants can also be effectively controlled with lopping shears or by individual plant treatment with approved herbicides. Herbicide application needs to be done carefully to avoid damage to nearby desirable trees such as oak, hackberry, or sumac. Biological control with goats is sometimes suggested as a means to control juniper, but this method is very risky and may have detrimental side effects. While goats can be stocked heavily enough to cause the damage or death of small juniper plants, their browsing of other, more desirable shrubs usually causes an overall decline in the abundance, vigor, and diversity of shrubs. The control of juniper is often suggested for the purpose of saving water. However, researchers have discovered that the control of brush, including juniper, especially in semiarid regions such as the Concho Valley, usually has no substantial off-site water benefits. In some cases, where the right geology exists, and where springs occurred in the past, control of Ashe juniper may result in some increased spring flow. If other woody plants such as oak begin to grow in these areas, spring flow will often decrease. The primary benefits to be gained are increased grass production, improved plant diversity, and better visibility for the management of livestock.

Fruit—Red berries, when present, indicate a female plant. This is the resprouter, as compared to Ashe juniper (*Juniperus ashei*).

Growth Form—Redberry juniper usually appears more shrublike and multistemmed, whereas Ashe juniper typically grows with a straighter stem or trunk and is more treelike.

# Redberry juniper *(Juniperus pinchotii)*
## CUPRESSACEAE—CYPRESS FAMILY

**Characteristics:** Native; evergreen.

**Growth Form:** Large evergreen shrub; height 12–18 feet.

**Reproduction:** Seeds.

**General Description:** This is by far the most common juniper in the Concho Valley. The difference between this and Ashe juniper is not always easy to determine when berries are not present. The leaves of redberry juniper usually have faint white resin deposits visible, although this material may get temporarily washed off by rainfall. This species is almost always multistemmed, whereas Ashe juniper often has a single trunk. Redberry juniper has a much sparser canopy compared to that of Ashe juniper. This has important implications for the amount of grass that will grow under juniper. Redberry is a lighter green compared to the dark green of Ashe juniper. Other common names: redberry cedar, cedar.

**Livestock and Wildlife Value:** This plant is often underrated for its value to livestock and wildlife. Goats, white-tailed deer, and sheep browse the leaves, especially in winter when other green vegetation is scarce. The browse is not preferred over that of many other shrubs, but it is widely available in large quantity. During a dry winter juniper often makes up a large part of the diet of goats and white-tailed deer. The berries are extremely valuable to many species of birds, but this is a mixed blessing since birds are also responsible for spreading juniper seeds in their droppings. Small mammals also readily consume the berries. A study conducted many years ago in Sterling County indicated that jackrabbits are responsible for much of the seed dissemination of redberry juniper. Juniper also provides important cover for wildlife and protection for livestock during severe weather.

**Management:** The abundance of juniper has increased markedly in the region over the last 50 to 100 years. Control of excess juniper has been a never-ending goal of many ranchers for

**Allergy Season**—The white haze in this picture is actually the pollen from the male plant, which easily disperses in the wind. This pollen triggers inhalant allergies in many people. Photos by George Clendenin, USDA-NRCS.

Grubbing or excavating the entire underground basal stump is necessary to kill redberry juniper (*Juniperus pinchotii*), since it is a resprouter. Photo by Mark Meyer, USDA-NRCS.

several generations. High densities of redberry juniper interfere with livestock ranching. Thick juniper makes it difficult if not impossible to see livestock and to gather them, check on them, or feed them. Dense juniper also limits the growth of other, desirable plants. Those plants may exist under the canopy of juniper but may be stunted and unproductive. Control of redberry juniper is more difficult than control of Ashe juniper, since redberry resprouts from the stump. Grubbing or excavation of the underground basal stump is necessary to kill the plant. Fire can be used to kill seedlings and small plants less than about 12 inches tall. Small plants can also be effectively treated by spot application of approved herbicides. Herbicide application needs to be done carefully to avoid damage to nearby desirable trees such as oak, hackberry, or sumac. Once a plant reaches a height above 12 inches, fire will only top-kill it and the stump will resprout. Biological control with goats is sometimes suggested as a means to control juniper, but this method is very risky and may have detrimental side effects. While goats can be stocked heavily enough to cause damage or death of small juniper plants, their browsing of other, more desirable shrubs usually causes an overall decline in the abundance, vigor, and diversity of shrubs. The control of juniper is often suggested for the purpose of saving water. However, researchers have discovered that the control of brush, including juniper, especially in semiarid regions such as the Concho Valley, usually has no substantial off-site water benefits. Redberry juniper grows more in the drier parts of the Concho Valley where springs are not common. Control of brush in these areas should not be expected to result in an increase in water supplies. The benefits to be gained are increased grass production, improved plant diversity, and better visibility for the management of livestock. Because the canopy of this plant is sparse, grass often grows in abundance under redberry juniper. Here, grass and other plants find cooler temperatures, rich soil, and improved moisture conditions.

Edible Fruits—The fruits when fully ripe (black) have a pleasing sweet flavor. Ripe persimmon berries have been used to produce jams, puddings, cream pies, and quick breads (Tull 1987). Pick only when ripe, in early summer or late fall; the unripe fruits are not a pleasant taste experience! Leaves are oval, with the edges curled downward.

Bark—The smooth, white or light gray bark is distinctive. Photos by George Clendenin, USDA-NRCS.

# Texas persimmon *(Diospyros texana)*
## EBENACEAE—PERSIMMON OR EBONY FAMILY

**Characteristics:** Native; deciduous.

**Growth Form:** Medium-sized shrub; height 6–12 feet.

**Reproduction:** Seeds and the sprouting of severed roots.

**General Description:** The most identifiable feature of this shrub is the distinctive, smooth, white or light gray bark. The female plants produce large, juicy black fruits in late summer. The sweet black persimmons are ¾–1 inch in diameter. The wood is extremely hard and heavy. Other common names: Mexican persimmon, black persimmon.

**Livestock and Wildlife Value:** Browsed by goats and white-tailed deer, but not preferred over other shrubs. The ripe fruits are relished by all classes of livestock and by white-tailed deer, turkeys, and small mammals. The extreme overconsumption of ripe fruits can cause animal health problems.

**Management:** This plant rarely gets thick enough in the region to justify control. It is normally a minor component of the landscape. If the central trunk is grubbed with mechanical equipment, the severed lateral roots will often sprout new bushes. Fire will sometimes top-kill the plants, which will resprout from the stump.

**Toxic Agent:** Currently unknown.

**Toxic Description:** Observed poisonings have not led to death but rather resulted in poor performance. See *Toxic Plants of Texas* (Hart et al. 2010) for livestock clinical signs and more information.

**Other Uses:** Besides having edible berries, Texas persimmon has very hard wood that works well for tools and wood sculpture. The berries have also been used to make dyes since early pioneering times (Tull 1987).

Photos by George Clendenin, USDA-NRCS.

# Ephedra *(Ephedra antisyphilitica)*
## EPHEDRACEAE—MORMON TEA FAMILY

**Characteristics:** Native; evergreen.

**Growth Form:** Low evergreen shrub; height 2–6 feet and of equal or greater diameter.

**Reproduction:** Seeds. Dioecious (pollen- and seed-producing cones on separate plants).

**General Description:** Shrub lacking leaves and appearing as a mass of green sticks. Female plants produce seeds inside fleshy red berries in spring. There is another species of ephedra in the Concho Valley area: vine ephedra (*E. pedunculata*), which is sometimes seen growing high in a tree or climbing up a fence post. Other common names: Mormon tea, clapweed, joint-fir, popote, cañatilla.

**Livestock and Wildlife Value:** Ephedra is browsed year-round by goats, sheep, and white-tailed deer but is of greatest value in winter when it may be one of the few green plants. The young tender joints contain about 12 percent crude protein (Everitt and Drawe 1993). Fruits and seeds are sometimes eaten by quail and other birds.

**Management:** This plant is best maintained by conservative grazing management and proper control of white-tailed deer numbers. It responds very positively to fire with vigorous new resprouts.

**Myths and Other Uses:** Mormon tea was one of many plants that were mistakenly thought to cure syphilis before the days of antibiotics (Tull 1987). American Mormon tea was also erroneously thought to contain high doses of ephedrine (thus the name "ephedra"), although its Chinese relative, ma huang, is used as a source for this powerful drug (Moore 2003). The Texas species are reported to contain only minute amounts of ephedrine, if any (Tull 1987; Moore 2003). The green branches of ephedra, however, can be used to make a tea. Ephedra has been found in medicine bundles dating back a thousand years or more (Moore 2003).

Flowers and Leaves—Flower clusters are elongated, not round as in other species of catclaw, with cream-colored flowers. There are 3–7 pairs of leaflets per compound leaf. The leaves and leaflets are smaller than those of Roemer's acacia.

Bean Pods—Pods are large, flat, and often twisted. Photos by George Clendenin, USDA-NRCS.

# Catclaw acacia *(Acacia greggii)*
## FABACEAE — LEGUME OR BEAN FAMILY

**Characteristics:** Native; deciduous.

**Growth Form:** Medium to large shrub or small tree. Usually grows as a multistemmed shrub to a height of 6–12 feet; sometimes grows as an impressive single-trunked gnarly tree up to 20 feet tall.

**Reproduction:** Seeds.

**General Description:** This is one of the 4 species in the region that can have recurved "catclaw" thorns (along with Roemer's acacia, catclaw mimosa, and fragrant mimosa). Thorns are often absent or sparse on this species in this region, although in other regions the thorns can be numerous. Other common name: uña de gato.

**Livestock and Wildlife Value:** Browsed by goats and white-tailed deer, but not preferred. Low, dense shrubs provide good cover for songbirds and good loafing cover for quail. Seeds are sometimes eaten by quail. Flowers are an important source of nectar for bees and other pollinating insects.

**Management:** This species is not abundant enough in the region to be considered a brush problem. It is usually considered desirable and is intentionally retained when conducting brush control.

Flowers—Flower clusters are round and cream colored. Photos by George Clendenin, USDA-NRCS.

Leaves—There are 4–8 pairs of leaflets per compound leaf. The leaves and leaflets of this species are larger than those of any other catclaw and the new stems have a distinctive reddish color.

# Roemer's acacia *(Acacia roemeriana)*
## FABACEAE—LEGUME OR BEAN FAMILY

**Characteristics:** Native; deciduous.

**Growth Form:** Ranges in size from a small shrub to a small tree.

**Reproduction:** Seeds.

**General Description:** This is one of the 4 species in the region that can have recurved "catclaw" thorns. Roemer's acacia grows in shallow, rocky soil, often in canyons, rimrock, or headers. Other common name: catclaw.

**Livestock and Wildlife Value:** Provides good browse value to goats, sheep, and white-tailed deer. This is the most preferred of the 4 species of catclaw. Seeds may be eaten by birds.

**Management:** This desirable species should be retained during brush-control operations. Careful attention to goat and white-tailed deer numbers will help maintain this shrub.

Leaves are pinnately compound, but with very tiny leaflets!

In Full Bloom—With flowers, this is a very beautiful plant. Without flowers, it is an easy-to-miss, very small shrub. Our dalea species are unarmed, meaning they do not have thorns. Photos by George Clendenin, USDA-NRCS.

# Feather dalea *(Dalea formosa)*
## FABACEAE—LEGUME OR BEAN FAMILY

**Characteristics:** Native; deciduous.

**Growth Form:** Very small shrub; height 12–24 inches.

**Reproduction:** Seeds.

**General Description:** Flowers are rich purple and yellow, with a feathery interior. When in full bloom, this is a beautiful plant. Feather dalea grows on very shallow soils. Leaves have a peculiar strong odor when crushed.

**Livestock and Wildlife Value:** Browsed to a limited extent by goats, sheep, and white-tailed deer.

**Management:** This small shrub requires no special management, but it is susceptible to some herbicides used for brush control. Heavy browsing on this plant is usually a sign of too many goats or white-tailed deer.

Pretty Flowers, Nasty Thorns—This species of catclaw has the most rigid and sharp thorns of the 4 species. Thorns are formed in matching pairs along the stem. Photo by Steve Nelle, USDA-NRCS.

# Catclaw mimosa
*(Mimosa aculeaticarpa* var. *biuncifera)*
## FABACEAE—LEGUME OR BEAN FAMILY

**Characteristics:** Native; deciduous.

**Growth Form:** Scraggly shrub; height 3–6 feet, often twice this in width.

**Reproduction:** Seeds.

**General Description:** The sparse flower clusters are white or sometimes faintly pink. New twigs often grow in zigzag fashion. Mature bean pods are small and grow in clusters. Bean pods often have numerous, conspicuous, recurved prickles on the margins. This plant grows in very shallow, rocky soils.

**Livestock and Wildlife Value:** Occasionally browsed by goats and white-tailed deer but not considered a desirable browse plant. Seeds are eaten by birds.

**Management:** This plant sometimes becomes thick enough to be a nuisance. Lambs and kid goats with long hair can become entangled in the thorny branches, where they can die from lack of water or food or exposure to predators or the elements. Catclaw mimosa can be seriously injured by fire but will usually resprout. Its most important ecological role is as a valuable protective nursery plant. Its network of spiny branches provides physical protection to other plants that grow inside it. Desirable grasses, forbs, and woody plants can often get established in the protection of catclaw mimosa. Under good grazing management, these desirable plants will be able to spread from these protected locations.

In Full Bloom—The distinctive bright pink, yellow-tipped, fragrant flower clusters of this species resemble a powder puff. Photos by George Clendenin, USDA-NRCS.

# Fragrant mimosa *(Mimosa borealis)*
## FABACEAE—LEGUME OR BEAN FAMILY

**Characteristics:** Native; deciduous.

**Growth Form:** Small, sparse shrub; height 3–6 feet.

**Reproduction:** Seeds.

**General Description:** This very small shrub goes unnoticed most of the time until it flowers. When in full flower, it becomes a beautiful specimen, at which point you may notice how many of these plants you actually have in your pasture! As with many of our drought-hardy range plants, these shrubs need good winter and spring precipitation to really show their colors. This is one of the 4 woody species that can have "catclaw" thorns. Mature bean pods often have no prickles on the margin, though sometimes they have an occasional one.

**Livestock and Wildlife Value:** Lightly browsed by goats and white-tailed deer but not preferred. Seeds are eaten by birds.

**Management:** This species never becomes abundant enough to justify brush control. No specific management is recommended.

**Bipinnately Compound Leaves**—Mesquite has this distinctive compound leaf structure. Flowers appear in drooping, yellowish-white catkins.

**Mesquite Beans**—Bean crops, as a part of a balanced forage diet, can provide a windfall of nutrition in the hottest part of summer; however, bean crops are not reliable. Photos by George Clendenin, USDA-NRCS.

# Honey mesquite *(Prosopis glandulosa)*
## FABACEAE — LEGUME OR BEAN FAMILY

**Characteristics:** Native; deciduous.

**Growth Form:** Growth form varies considerably, ranging from dense thickets of multistemmed brush to large impressive trees up to 40 feet tall.

**Reproduction and Dispersal:** Seeds are produced in bean pods. Livestock and wildlife eat the pods while still on the tree or after they have fallen to the ground. Cattle readily consume, digest, and disperse viable mesquite seeds. Goats eat more seeds by body weight than cattle, but goats and sheep do not disperse viable seeds as cattle do. Even without livestock grazing, mesquite beans may not stay on the ground very long because various wildlife species may consume them (Kneuper, Scott, and Pinchak 2003). Most mesquite seeds are destroyed by weevils, but plenty survive intact to perpetuate the species.

**General Description:** The most well-known and common woody plant in the region. Although mesquite is without doubt a native plant that belongs here, its abundance has increased dramatically over the past 100 years. Mesquite developed a reputation many years ago of using a lot of water. Prevailing popular opinion held that removal of mesquite would markedly improve water supplies. Scientific studies have since indicated very small if any water benefits from mesquite control in semiarid climates (Wilcox and Huang 2010; Wilcox 2002). The primary benefits of mesquite control include improved visibility in pastures for the management of livestock and increased forage production.

**Livestock and Wildlife Value:** Mesquite is generally underappreciated as a valuable plant for livestock and wildlife. Leaves are regularly browsed by livestock and white-tailed deer but are not preferred. Dead frostbitten leaves are eaten in the fall and winter. Mesquite beans are relished by all kinds of livestock and by white-tailed deer and are of high nutritional value, but overconsumption for a prolonged period (60 days or more) can

While some mesquite can be desirable for livestock and wildlife habitat, an overabundance can reduce available forage, as well as make it difficult to manage livestock. Photo by Mark Meyer, USDA-NRCS.

Mechanical removal of mesquite is an effective way to manage its populations. Excavators are commonly used in the Concho Valley and are effective in rocky soil types as well as deep soils. Photo by Dee Ann Littlefield, USDA-NRCS.

cause serious health problems for cattle and goats. Mesquite provides adequate cover on millions of acres to sustain diverse wildlife populations. Its thin canopy and partial shade create favorable conditions for many desirable plants to grow under it. Being a legume, mesquite is able to enrich the soil by converting atmospheric nitrogen into soil nitrogen. Many astute ranchers have noticed the abundant Texas wintergrass that often grows in association with mesquite. This is a good ecological combination: a deciduous tree that enriches the soil and a cool-season grass that grows in the winter when mesquite is dormant. Despite its considerable value, mesquite has become a serious brush problem on most ranches. Dense mesquite inhibits visibility, making it extremely difficult to manage livestock properly, as well as reducing forage production because of competition for moisture.

**Management:** The suppression or management of mesquite has become a priority for many ranches in the region. Methods of control include mechanical and chemical for suppression, and prescribed fire for manipulation. The goal is generally to reduce the density of mesquite in some areas while retaining it for wildlife habitat in others. Some ranchers who have limited interest in wildlife sometimes desire complete control of mesquite.

MECHANICAL AND CHEMICAL CONTROL: Mechanical methods that uproot the underground base of the trunk are necessary since mesquite is a strong stump sprouter. Chemical methods include foliar sprays, which generally must be applied in midsummer; basal treatment; and cut stump treatment.

PRESCRIBED FIRE: Fire is ineffective in killing mesquite but can be used to temporarily reduce the canopy. The best use of fire in mesquite pastures is to "prune" the lower canopy with a cool-season fire, but not to burn the entire top. This will allow the tree to persist but will not encourage resprouting from the stump.

**Acorns**—These are the largest of any oak in the world, measuring up to 2 inches in diameter including the cup.

**Leaves**—The leaves are large and lobed, usually with a glossy top surface (top photo) and a pale underside (bottom photo). The top lobe of the leaf is always the largest.

**Along the Banks**—On the South Concho River in Christoval, this bur oak stands majestically. If you like to collect the acorns, you need to be quick, or the squirrels and other wildlife will get to them first! Photos by George Clendenin, USDA-NRCS.

# Bur oak *(Quercus macrocarpa)*
## FAGACEAE—OAK FAMILY

**Characteristics:** Native; deciduous.

**Growth Form:** Large, stately tree; height to 80 feet.

**Reproduction:** Seeds.

**General Description:** Bur oak and pecan are rivals for the tallest trees in the region. Within the Concho Valley, bur oak grows only along the South Concho River, which is the most western occurrence of the species in Texas. It also occurs along the San Saba River in western Menard County. Leaves are large, with several deep lobes.

**Livestock and Wildlife Value:** The acorns are consumed by livestock, white-tailed deer, turkeys, and squirrels. Livestock and white-tailed deer browse on young trees. Trees provide good roosting habitat for turkeys, especially when mixed with other large trees such as pecan and live oak.

**Management:** Landowners revere bur oak and protect it from cutting and disturbance. Natural reproduction of bur oak is generally inadequate to maintain desirable populations as older trees die. Seedlings are especially vulnerable to browsing by livestock and white-tailed deer. Careful attention to grazing management and white-tailed deer numbers is necessary to allow more trees to establish in riparian areas. Bur oak seedlings and saplings are commercially available and are commonly planted in parks and yards.

**Toxic Agent:** Gallotannins.

**Toxic Description:** All species of oak (*Quercus*) should be considered potentially toxic, but they seldom pose a problem to livestock in this region.

Photos by George Clendenin, USDA-NRCS.

# Mohr shin oak *(Quercus mohriana)*
## FAGACEAE—OAK FAMILY

**Characteristics:** Native; deciduous.

**Growth Form:** Varies a great deal; often grows in colonies of low, dense, shrubby bushes; also grows in mottes of small trees up to 16 feet tall.

**Reproduction:** Root sprouting and acorns.

**General Description:** This is one of the 3 shrubby species of oak in the region that are referred to as shin oak (along with white and Vasey). Bark of all shin oaks is light gray and scaly. This and the other shin oaks typically grow in shallow limestone soils in canyons, headers, and rimrock areas, but they can also grow in the bottoms of narrow canyons. Mohr shin oak is more commonly found in the northern and western Concho Valley and is commonly seen mixed with white shin oak along the divide between the Concho and Colorado watersheds.

**Livestock and Wildlife Value:** Browsed by livestock (especially goats) and white-tailed deer but not as desirable as white shin oak. The nutritional quality of the browse is marginal, but the abundance of available browse is often high. Acorns are consumed by livestock, white-tailed deer, turkeys, and many other animals. It should be noted that the buds, fresh tender leaves, and flowers can be toxic to livestock for 2 or 3 weeks in early spring. During this time, the tannin content of the new growth is very high and can poison cattle, sheep, or goats. Shin oak thickets provide exceptional cover for white-tailed deer and a large variety of other wildlife.

**Management:** Shin oak is usually considered a desirable woody plant and is protected during brush control operations. Much of the browse production of shin oak can grow out of the reach of goats and white-tailed deer as the plants grow old. In these cases, mechanical methods or fire can be used to knock down the top growth and stimulate the production of root sprouts. To reduce or prevent toxicity problems, move livestock out of shin oak pastures during the period when new growth appears in spring.

**Leaves**—The primary identifying feature of this species is the distinctive grayish-green color of the leaves. The underside of the leaves is covered by dense, fine, silvery fuzz. Leaf shape is variable but is usually oval, with or without several shallow, rounded or pointed lobes.

**Toxic Agent:** Gallotannins.

**Toxic Description:** These compounds are toxic to cattle, sheep, goats, horses, and dogs. Usually the toxin is more likely concentrated in young leaves, stems, buds, and flowers. Toxicity is more likely during 2 or 3 weeks in early spring. Mature leaves are not toxic. Poisoning can occur when as little as 6 percent of an animal's body weight is consumed. See *Toxic Plants of Texas* (Hart et al. 2010) for livestock clinical signs and more information.

Photo by Steve Nelle, USDA-NRCS.

# Vasey shin oak *(Quercus pungens* var. *vaseyana)*
(Sy = *Q. pungens*)
**FAGACEAE—OAK FAMILY**

**Characteristics:** Native; deciduous.

**Growth Form:** Varies; may grow in low, dense colonies of shrubby bushes or as small trees up to 12 feet tall.

**Reproduction:** Root sprouting and acorns.

**General Description:** This is one of the 3 shrubby species of oak in the region that are generically referred to as shin oak. The primary identifying feature of this species is the small, rough-textured leaves with shallow, sharp-pointed lobes. The underside of the leaves is covered by stiff hairs. The bark of all shin oaks is light gray and scaly. This and the other shin oaks typically grow in shallow limestone soils in canyons, headers, and rimrock areas, but they can also grow in the bottoms of narrow canyons. Vasey oak is found in the southern and western part of the region and is sometimes found mixed with white shin oak. Other common name: sandpaper oak.

**Livestock and Wildlife Value:** Browsed by livestock (especially goats) and white-tailed deer, but less palatable because of the coarse leaves. The nutritional quality of the browse is marginal, but the abundance of available browse is often high. Acorns are consumed by livestock, white-tailed deer, turkeys, and many other animals. As with all of the shin oaks, the buds, fresh tender leaves, and flowers can be toxic to livestock for 2 or 3 weeks in early spring. During this time, the tannin content of the new growth is very high and can poison cattle, sheep, or goats. Shin oak thickets provide exceptional cover for white-tailed deer and a large variety of other wildlife.

**Management:** Vasey oak is considered a desirable woody plant and is usually protected during brush-control operations. Much of the browse production can grow out of the reach of goats and white-tailed deer as the plants grow old. In these cases, mechanical methods or fire can be used to knock down the top growth and stimulate the production of root sprouts. To reduce

Photo by Steve Nelle, USDA-NRCS.

Photo by Steve Nelle, USDA-NRCS.

or prevent toxicity problems, move livestock out of shin oak pastures during the period when new growth appears in spring.

**Toxic Agent:** Gallotannins.

**Toxic Description:** These compounds are toxic to cattle, sheep, goats, horses, and dogs. Usually the toxin is more likely concentrated in young leaves, stems, buds, and flowers. Toxicity is more likely during 2 or 3 weeks in early spring. Mature leaves are not toxic. Poisoning can occur when as little as 6 percent of an animal's body weight is consumed. See *Toxic Plants of Texas* (Hart et al. 2010) for livestock clinical signs and more information.

Bark of all shin oaks is light gray and scaly.  Photos by George Clendenin, USDA-NRCS.

390

# White shin oak *(Quercus sinuata* var. *breviloba)*
## FAGACEAE—OAK FAMILY

**Characteristics:** Native; deciduous.

**Growth Form:** Varies a great deal; often grows in colonies of low, dense, shrubby bushes; also grows as mottes of small trees up to 16 feet tall.

**Reproduction:** Root sprouting and acorns.

**General Description:** This is one of the 3 shrubby species of oak in the region that are referred to as shin oak. Leaves of this species are variable but usually have several shallow, rounded lobes. Bark of all shin oaks is light gray and scaly. This and the other shin oaks typically grow in shallow limestone soils in canyons, headers, and rimrock areas, but they can also grow in the bottoms of narrow canyons. White shin oak is more commonly found in the eastern and southern Concho Valley.

**Livestock and Wildlife Value:** Frequently browsed by livestock (especially goats) and white-tailed deer. The nutritional quality of the browse is marginal, but the abundance of available browse is often high. Acorns are consumed by livestock, white-tailed deer, turkeys, and many other animals. Shin oak thickets provide exceptional cover for white-tailed deer and a large variety of other wildlife. The drooping flower spikes, called catkins, attract a host of tiny insects, which in turn provide a rich food source for many small songbirds in late winter or early spring when other food is scarce.

**Management:** Shin oak is usually considered a desirable woody plant and is protected during brush-control operations. Much of the browse production of shin oak can grow out of the reach of goats and white-tailed deer as the plants grow old. In these cases, mechanical methods or fire can be used to knock down the top growth and stimulate the production of root sprouts. To reduce or prevent toxicity problems, move livestock out of shin oak pastures during the period when new growth appears in spring.

**Toxic Agent:** Gallotannins.

**Toxic Description:** These compounds are toxic to cattle, sheep, goats, horses, and dogs. Usually the toxin is more likely concentrated in young leaves, stems, buds, and flowers. Toxicity is more likely during 2 or 3 weeks in early spring. Mature leaves are not toxic. Poisoning can occur when as little as 6 percent of an animal's body weight is consumed. See *Toxic Plants of Texas* (Hart et al. 2010) for livestock clinical signs and more information.

Photo by Steve Nelle, USDA-NRCS.

Flowering—The catkins of flowers may be easy to miss. Photo by George Clendenin, USDA-NRCS.

# Live oak *(Quercus virginiana)* (Sy = *Q. fusiformis*)
## FAGACEAE—OAK FAMILY

**Characteristics:** Native; evergreen (late deciduous).

**Growth Form:** Varies from thickets of scrubby root sprouts to clumps or mottes of medium-sized trees, to very large single trees up to 60 feet tall.

**Reproduction:** Seeds and root sprouts.

**General Description:** Live oak is one of the most well-loved and easily recognized trees in the region. Although often described as evergreen, it does lose its leaves every year for a brief period in late winter. Leaf shape varies greatly, even on the same tree. The leaves of root sprouts may look different from leaves on a mature tree. All 3 common growth forms can be found in close proximity, but mottes of 10–20 medium-sized trees sharing the same root system are the most common in the Concho Valley.

**Livestock and Wildlife Value:** Live oak is extensively browsed by livestock and white-tailed deer, although the nutritional value of the browse is marginal. Acorns are eaten in large quantities by goats, sheep, white-tailed deer, turkeys, larger birds, squirrels, and other small mammals. Acorns often provide a windfall of nutrition that helps animals go into winter in good condition, but acorn production is not reliable. Live oak also provides excellent cover for a wide variety of wildlife.

**Management:** Live oaks are one of the most prized trees and are usually protected by landowners, who value their appearance and character and their value to livestock and wildlife. High-intensity fires (either wildfire or human-caused), especially fires that ignite the canopy, can kill these trees. However, well-planned prescribed burns clean up the understory of these trees, thereby helping preserve them from wildfire. Landowners wishing to protect large trees from fire damage should be careful to remove slash piles and ladder fuels such as other living shrubs and brush away from desirable trees. This practice also helps preserve these majestic trees during a wildfire. Prescribed burning can also be used to greatly improve the availability

Photo by George Clendenin, USDA-NRCS.

of browse since it stimulates root sprouting and increases the nutritional value of browse for several months following a fire. Oak wilt, a fungal disease, is a potential threat to live oak in the region. Landowners should be very careful when bringing in firewood from other regions where oak wilt is present.

**Toxic Agent:** Gallotannins.

**Toxic Description:** All species of oak (*Quercus*) should be considered potentially toxic, but live oak seldom poses a problem to livestock in this region.

Catkins—These drooping flower clusters are called catkins. The catkins and leaves show up at the same time. Pecans have both flower sexes on the same plant. The catkins are the male flowers, which means the pollen-producing ones. This is significant for people who suffer from pecan inhalant allergies.

Pecan Trees—These grow native, especially in stream bottoms, but are also frequently used in native plantings for their shade and nut production. Photos by George Clendenin, USDA-NRCS.

# Pecan *(Carya illinoinensis)*
## JUGLANDACEAE — WALNUT FAMILY

**Characteristics:** Native; deciduous.

**Growth Form:** Large tree up to 80 feet tall.

**Reproduction:** Seeds.

**General Description:** Pecan is the State Tree of Texas and no stranger to the residents of the Concho Valley. Most of the pecan trees in the region have been planted in towns, farmsteads, and orchards, but there are also lots of native pecans in the river bottoms and along major creeks.

**Livestock and Wildlife Value:** Nuts are eaten by livestock, white-tailed deer, feral hogs, turkeys, large songbirds, squirrels, and other small mammals. Groves of large pecans create ideal roosting habitat for turkeys, which may congregate there by the hundreds.

**Management:** Pecan trees are highly regarded in the region, and all landowners protect them. Of special importance is the natural reproduction and recruitment of new pecan trees. In some areas there are many older trees but few if any young trees; sometimes the old trees are dying of natural causes, especially drought, but no young trees are there to replace them. Very conservative grazing in riparian areas will help promote the successful reproduction of pecans and other trees. Many landowners graft improved varieties onto their native pecan trees and provide additional management to increase the production of nuts, which are sold as a cash crop.

**Mature Fruit**—The walnuts are the seeds inside this fleshy fruit!

**Bark**—Typical rough bark of little walnut.

**Leaves**—All leaves in the Juglandaceae family are alternate and pinnately compound. The mature leaflets of little walnut are lightly serrated on the edges. The end (terminal) leaf in *Juglans* species is always smaller than the middle lateral leaflets. This is not true for *Carya* species, which means this rule does not apply to pecans.

**Leaf Cluster**—Notice how the individual leaflets turn slightly so they remain roughly parallel with the ground surface, even if the branch is drooping. Photos by George Clendenin, USDA-NRCS.

# Little walnut *(Juglans microcarpa)*
## JUGLANDACEAE — WALNUT FAMILY

**Characteristics:** Native; deciduous.

**Growth Form:** Small tree; height to 20 feet.

**Reproduction:** Seeds.

**General Description:** Grows in creek bottoms and riparian areas. This species is sometimes difficult to distinguish from smaller pecan trees. Flowers form in early spring in long drooping clusters called catkins. After the flowers are pollinated by the wind, fruits begin to form in summer. Husks are green, turning black at maturity. Nuts are about ¾ inch in diameter.

**Livestock and Wildlife Value:** Browsed by goats and white-tailed deer but not considered preferred browse. The walnuts are extremely hard and the extraction of the meat is difficult; provides a limited food source for squirrels.

**Management:** This plant normally needs no specific management. Care should be taken to retain this plant during brush control in creek or river bottom areas. These trees, along with other riparian trees, are important in helping hold creek banks together during flood events. The massive root systems help stabilize the channel and resist erosion.

ID Tip—Leaves have very faint serrations along the edge, and twigs often have small white spots, which can aid in identification.

Elbows—Plant is distinguished by the presence of oppositely paired matching twigs and opposite leaves. The oppositely paired twigs almost form a 90° angle to the stem; thus the common name "elbowbush."

Growth Form—Elbowbush often occurs in thickets under the canopy of live oak. Photos by George Clendenin, USDA-NRCS.

# Elbowbush *(Forestiera pubescens)*
## OLEACEAE — OLIVE FAMILY

**Characteristics:** Native; deciduous.

**Growth Form:** Scraggly shrub; height 6–12 feet.

**Reproduction:** Seeds, and occasionally the rooting of low stems that touch the ground.

**General Description:** Plant flowers before leaves form. This is usually the first shrub to produce green leaves, usually in late February or early March. It is also the first to begin losing its leaves in fall. Clusters of small purple berries form in early summer.

**Livestock and Wildlife Value:** Elbowbush is a desirable browse for goats, sheep, and white-tailed deer. Berries are eaten by birds. Dense thickets of elbowbush provide good cover for a large variety of wildlife.

**Management:** This shrub should be saved during brush-control operations. Proper management of white-tailed deer and goat numbers is important to keep this plant in good productive condition.

Leaves—Leaves are extremely small and crowded together in clusters on the stem.
Photo by George Clendenin, USDA-NRCS.

Photo by Steve Nelle, USDA-NRCS.

Photo by George Clendenin, USDA-NRCS.

# Javelina bush *(Condalia ericoides)*
(Sy = *Microrhamnus ericoides*)
## RHAMNACEAE—BUCKTHORN FAMILY

**Characteristics:** Native; evergreen.

**Growth Form:** Low, spiny, evergreen shrub; height 1–3 feet.

**Reproduction:** Seeds.

**General Description:** This is a western-southwestern shrub. It grows in the western and northwestern part of the Concho Valley, mostly on shallow soils. The leaves are extremely small and crowded together in clusters. Each twig terminates in a sharp spine. Flowers are inconspicuous. Berries are large, reddish brown to black, and pointed at the tip.

**Livestock and Wildlife Value:** Rarely browsed by livestock or wildlife. Berries are presumably eaten by birds or small mammals, but this has not been confirmed.

**Management:** This shrub requires no specific management.

Green condalia (*Condalia hookeri*) and lotebush (*Ziziphus obtusifolia*) are similar in that their branched tips end in thorns. Green condalia can be distinguished from lotebush by its bright green leaves, which are broader at the end and often tipped by a small point. Green condalia also does not have the waxy branches of lotebush. Photos by George Clendenin, USDA-NRCS.

# Green condalia *(Condalia hookeri)* (Sy = *C. obovata*)
## RHAMNACEAE—BUCKTHORN FAMILY

**Characteristics:** Native; deciduous to partially evergreen.

**Growth Form:** Low, spiny shrub; height 2–8 feet.

**Reproduction:** Seeds.

**General Description:** Green condalia (*Condalia hookeri*) is very similar to and often confused with lotebush (*Ziziphus obtusifolia*). They are both medium-height shrubs with small leaves, whose branch tips end in thorns. Although leaf shapes can be variable between the two, green condalia leaves can be distinguished by being bright green and broader toward the tip, and the tip also often has a small point. Green condalia does not have waxy branches like lotebush. Fruits of green condalia range from green to red to black depending on ripeness.

**Livestock and Wildlife Value:** Browsed to a limited extent by goats and white-tailed deer in spring when the new twigs are tender and soft. Berries are eaten by many species of birds. Plants provide cover and protection for birds and other small animals.

**Management:** No specific management is recommended for this plant. It does not become abundant enough to warrant any kind of control.

**Other Uses:** As a source of dye, green condalia yields outstanding colors of green gold and a rich gray (Tull 1987). The mature purple-black berries are edible and can be eaten raw. The berries were used during frontier days to make jellies (Diggs, Lipscomb, and O'Kennon 1999).

Lotebush leaves are usually a duller grayish green than the brighter leaves of green condalia. Lotebush can best be distinguished by its grayish-white waxy branches, which often have fine parallel growth cracks. Photo by Steve Nelle, USDA-NRCS.

# Lotebush *(Ziziphus obtusifolia)* (Sy = *Condalia obtusifolia; Rhamnus obtusifolia; Z. lycioides*)
## RHAMNACEAE—BUCKTHORN FAMILY

**Characteristics:** Native; deciduous.

**Growth Form:** Scraggly bush; height 5–10 feet.

**Reproduction:** Seeds.

**General Description:** Lotebush (*Ziziphus obtusifolia*) is very similar to and often confused with green condalia (*Condalia hookeri*). They are both medium-height shrubs with small leaves, whose branch tips end in thorns. Lotebush leaf shapes can vary from oblong to nearly linear, and they are usually a duller grayish green than the brighter leaves of green condalia. Lotebush can best be distinguished by the grayish-white waxy branches that often have fine parallel growth cracks. When present, fruits are usually dark purple to black.

**Livestock and Wildlife Value:** Occasionally browsed by goats, sheep, and white-tailed deer, when the new fresh twigs are soft in spring. Otherwise, a poor browse plant. Lotebush is highly valued as loafing cover for quail. This is the place where coveys rest during midday between feeding periods and find protection from the elements and hawks. The fruits are eaten by a few species of songbirds and also by gray fox and ringtail cat.

**Management:** Most landowners retain lotebush during brush-control operations. In some rare situations, lotebush can become abundant enough to justify thinning. Very hot fires will completely top-kill lotebush and destroy its cover value for many years. It sprouts back after fire and mechanical damage.

**Native American Uses:** The boiled fresh root of this plant makes a sudsy soap that was used by the Apaches for washing their hair and for treating a variety of scalp conditions. The Apaches and Comanches used it to bathe skin sores and irritations; they also used it to bathe the flanks and backs of injured horses (Moore 1989).

Photos by George Clendenin, USDA-NRCS.

# Buttonbush *(Cephalanthus occidentalis)*
## RUBIACEAE—COFFEE OR MADDER FAMILY

**Characteristics:** Native; deciduous.

**Growth Form:** Medium-sized shrub; height 6–12 feet.

**Reproduction:** Seeds.

**General Description:** This shrub grows in wet areas, especially near creeks and rivers. It may also grow in seepy areas or where there is a high water table. Leaves are large, shiny green, and often arranged in triplets. Flowers are in dense round white balls. Seedheads are compact round clusters of brown seeds. Other common name: button willow.

**Livestock and Wildlife Value:** Buttonbush is modestly browsed by goats and white-tailed deer. The flowers are heavily used by butterflies. The seeds are eaten by songbirds and waterfowl. Beavers commonly cut the branches and eat the bark of buttonbush.

**Management:** This plant is an important stabilizer of creek banks and riparian areas. The roots are especially well adapted to resisting erosion. Disturbances should be minimized in riparian areas to maintain dense combinations of trees, shrubs, grasses, and sedges. Grazing should be light and periodic, not continuous.

**Toxic Agent:** Unknown.

**Toxic Description:** Buttonbush has been found to be poisonous to cattle, yet it is rarely browsed by cattle so toxicity problems are rare. It apparently does not affect goats, which sometimes browse it heavily. See *Toxic Plants of Texas* (Hart et al. 2010) for livestock clinical signs and more information.

Fruits—These are reddish brown and wrinkled.

Growth Form—Tickle tongue can be a small shrub to a small spindly tree. Photos by George Clendenin, USDA-NRCS.

# Tickle tongue *(Zanthoxylum hirsutum)*
*(Sy = Z. carolinianum var. fruticosum; Z. clava-herculis var. fruticosum)*

**RUTACEAE — CITRUS FAMILY**

**Characteristics:** Native; deciduous.

**Growth Form:** Large thorny shrub or small tree; height 12–18 feet.

**Reproduction:** Seeds; new plants also sprout from severed roots.

**General Description:** Twigs and branches are very thorny. Leaves are composed of several smaller shiny green leaflets, with small thorns on the central stem. Leaves have a strong citrus smell when crushed. Nearly everyone raised in the country is familiar with the practice of chewing the young tender leaves or branches of this shrub, which causes a strong tingling sensation on the lips, tongue, and gums. Other common names: toothache tree, Hercules-club, prickly-ash.

**Livestock and Wildlife Value:** Tickle tongue is browsed by goats, sheep, and white-tailed deer but is not preferred. It is browsed mostly when other, more desirable plants are not available. Seeds are very commonly eaten by doves, quail, and other birds. It provides good loafing cover for quail and general screening cover for a variety of wildlife.

**Management:** This plant is sometimes targeted for brush control, although it seldom becomes abundant enough to justify it. If the central trunk is uprooted, the severed lateral roots often produce new plants. Prescribed burning can be used to reduce the canopy of this and most other rangeland shrubs. The nutritional quality of the browse is greatly enhanced following fire, and the resprouting stumps will often be heavily browsed for the first season after a burn.

Flowers are formed in long clusters called catkins. Photo by George Clendenin, USDA-NRCS.

Community—Black willow (right) is sharing a riparian community with willow baccharis (left). Photo by Steve Nelle, USDA-NRCS.

# Black willow *(Salix nigra)*
## SALICACEAE—WILLOW FAMILY

**Characteristics:** Native; deciduous.

**Growth Form:** Medium-sized deciduous tree; height to 40 feet.

**Reproduction:** Seeds and fresh cuttings.

**General Description:** Grows near creeks and rivers and other wet areas. Leaves are long and narrow, with very fine serrations. Flowers form in long clusters called catkins; seeds are surrounded by light cottony fluff and are blown in the wind.

**Livestock and Wildlife Value:** Readily browsed by livestock and white-tailed deer. Beavers, which are not uncommon in the Concho Valley, cut willow trees and consume the bark as food.

**Management:** Black willow is an important riparian tree with a strong binding root system that helps hold banks in place during flood events. Creek and river areas should be grazed very lightly or intermittently to promote healthy riparian vegetation, and white-tailed deer numbers should be managed. Herbicide use is discouraged in riparian areas; if necessary, use with great caution and restraint. Black willow is one of the few trees that can be planted from dormant cuttings. Branches from 1 to 4 inches in diameter can be cut in late winter and planted in suitable wet locations with the butt of the cutting in the water table. The poles will root in spring as the soil warms up.

**Other Uses:** *Salix* species contain salicin and salicylic acid in the bark. Salicylic acid is similar to the acetylsalicylic acid found in aspirin. Therefore native peoples and European pioneers chewed *Salix* twigs as a pain reliever. The plants have also been made into a tea and used for other medicinal reasons. *Salix* species have been used as landscape ornamentals, and the pliable branches have been used in basketry. They have also been cut for timber (Moore 2003; Diggs, Lipscomb, and O'Kennon 1999).

Fruit—Large clusters of opaque yellow to amber berries with a large black seed in the center begin to form in summer and mature in fall. Do not eat the berries! They contain toxic saponins, which can also irritate the skin if handled.

Bark—Bark is a distinctive gray color.

Compound Leaves—Leaf edges are smooth (mostly), unlike the slightly to strongly serrated edges of *Carya* (pecan) and *Juglans* (little walnut) leaves. Leaves are usually even-pinnately compound, which means they do not have a terminal leaf or an odd leaf at the tip, although this in itself will not lead to a positive ID.

Fall Color—Leaves turn a distinctive soft yellow in the fall. Photos by George Clendenin, USDA-NRCS.

# Western soapberry *(Sapindus saponaria* var. *drummondii)*
## SAPINDACEAE—SOAPBERRY FAMILY

**Characteristics:** Native; deciduous.

**Growth Form:** Medium-sized tree to 40 feet tall, or more commonly grows as a grove of smaller trees sharing the same root system. Sometimes seen growing as a thicket of root sprouts.

**Reproduction:** Seeds and root sprouts.

**General Description:** Grows mostly in or near draws or on the banks of creeks but can also be found away from creeks. Large clusters of tiny white flowers appear in late spring. This tree is sometimes confused with the exotic chinaberry tree (*Melia azedarach*), which has berries of similar size and color. Other common name: wild chinaberry (this name leads to some of the confusion with exotic chinaberry).

**Livestock and Wildlife Value:** Commonly browsed by livestock and white-tailed deer. Turkeys eat the berries and commonly use groves of soapberry as roost locations.

**Management:** Soapberry is highly valued and is usually protected during brush-control operations. It can be maintained or increased by careful management of livestock grazing and white-tailed deer numbers. Groves can be developed or encouraged by temporary fencing to encourage root sprouting. Root sprouts may form 50 feet or more from mother trees.

**Other Uses:** Villagers in Mexico use the berries for laundry soap; a high content of saponin (a toxic alkaloid) yields a cleansing lather. The berries also make a bright yellow dye for wool (Tull 1987) and when crushed have been used to poison fish (Diggs, Lipscomb, and O'Kennon 1999).

**Fruits—** These will eventually mature into black berries.

**Bark—** The distinctive bark forms fine vertical ridges. Photos by George Clendenin, USDA-NRCS.

**Flowers and Leaves—** Small white flowers form in clusters in early summer. Several leaves often grow in clusters from the same location on the twig.

# Spiny bumelia *(Sideroxylon lanuginosum)*
(Sy = *Bumelia lanuginosa*)
## SAPOTACEAE—CHICLE, SAPODILLA, OR SAPOTE FAMILY

**Characteristics:** Native; deciduous.

**Growth Form:** Medium-sized tree; height to 30 feet, or mottes of smaller trees connected by a common root system.

**Reproduction:** Seeds and root sprouts.

**General Description:** Common in creek bottoms, but also found in draws and on shallow hillsides and hilltops. Leaves have a rounded tip and a narrow, tapered base. Twigs often terminate in a spine. Juicy black berries containing a single large seed mature in late summer or early fall. Other common names: chittamwood, bumelia, woollybucket bumelia.

**Livestock and Wildlife Value:** Spiny bumelia provides excellent browse for livestock and white-tailed deer. Berries are excellent food for quail, turkeys, and many other birds and small mammals. Mottes provide good cover for a variety of wildlife. White-tailed and axis deer bucks frequently rub their antlers on these trees, which exposes the reddish inner bark. In extreme cases, the trunk is completely girdled, leading to the death of the tree.

**Management:** This is one of the most valuable plants in the region for wildlife and should be protected during brush-control operations. Careful attention to grazing management and white-tailed deer numbers will help maintain this plant. Bumelia responds favorably to fire with increased root sprouting.

**Other Uses:** Bumelia berries have been successfully used in making dark blue, blue-green, and gray-green dyes. The ripe, sweet black berries have been used in jelly recipes (Tull 1987). Caution: digestive disturbances and dizziness are also reported from eating berries in greater quantities (Diggs, Lipscomb, and O'Kennon 1999). This tree is also a good bee tree in terms of honey production (Diggs, Lipscomb, and O'Kennon 1999). The wood is very dense and hard and has been used in making tool handles and cabinetry (Vines 1960).

Photos by George Clendenin, USDA-NRCS.

Leaves—Leaves are long and narrow and often grow in clusters from the same location. Twigs and stems are often armed with short spines.

# Wolfberry *(Lycium berlandieri)*
## SOLANACEAE—NIGHTSHADE FAMILY

**Characteristics:** Native; deciduous.

**Growth Form:** Small to medium-sized sprawling shrub; height 3–6 feet.

**Reproduction:** Seeds; can also be propagated by semihardwood cuttings (Nokes 2001).

**General Description:** A desirable but unattractive shrub for our region. Flowers are small and white or lavender. Numerous small red berries are produced in summer or fall. Wolfberry will lose its leaves during midsummer drought and may appear dead, but leaves will reappear after rainfall. Leaves may be retained during winter. Other common name: Berlandier's wolfberry.

**Livestock and Wildlife Value:** Infrequently browsed by goats and white-tailed deer. The red berries are relished by quail, white-winged doves, turkeys, and many species of songbird. Thickets of wolfberry can provide good loafing cover for quail and protective cover for songbirds.

**Management:** Wolfberry is very sensitive to some herbicides commonly used for brush control. Those interested in wildlife habitat should protect this shrub.

Invasive—Salt cedar is a perfect example of the danger of introducing exotic plants. Introduced from Europe many years ago for erosion control, this shrub has now become widely established in West Texas and the western United States, where it is considered invasive and undesirable. Photos by George Clendenin, USDA-NRCS.

# Salt cedar *(Tamarix gallica)*
## TAMARICACEAE—SALT CEDAR FAMILY

**Characteristics:** Introduced; deciduous.

**Growth Form:** Shrub or small tree; height to 25 feet.

**Reproduction:** Seeds and root sprouts.

**General Description:** Salt cedar was introduced from Europe and is now a Texas state-listed noxious plant. Salt cedar grows mostly in wet areas such as creek bottoms or the margins of lakes and reservoirs. Reddish twigs are long, thin, and flexible. Leaves are thin and delicate and somewhat resemble the leaves of juniper. Flowers form in long pink spikes. Seeds are minute.

**Livestock and Wildlife Value:** Browsed to some degree by goats and white-tailed deer. Has some value as cover for wildlife.

**Management:** Salt cedar can form dense thickets that monopolize resources and inhibit the growth of native riparian plants. Control of this shrub is considered a high priority by many landowners and conservationists. Salt cedar can be effectively controlled by herbicides.

Bark—The bark of hackberry is gray, often with distinctive corky warts or a general rough appearance.

Leaves—Leaves have an obvious network of veins, which are especially visible on the lower surface. The upper surface of the leaves is extremely rough to the touch, which helps identify this plant.

Browse Lines—Unfortunately, browse lines are common on hackberry trees, which are highly desirable for goats and deer alike and therefore also act as an indicator of too many animals. Photos by George Clendenin, USDA-NRCS.

# Netleaf hackberry *(Celtis laevigata var. reticulata)*
ULMACEAE — ELM FAMILY

**Characteristics:** Native; deciduous.

**Growth Form:** Medium-sized tree; height to 25 feet.

**Reproduction:** Seeds.

**General Description:** Hackberry, although mistakenly regarded by some as undesirable, is actually highly desirable browse for goats and white-tailed deer in our region. Fruits are dry, reddish or brown berries that mature in fall and may hang on the tree for several months.

**Livestock and Wildlife Value:** Highly preferred browse, consumed by livestock and white-tailed deer. Berries are relished by a large number of birds and small mammals. Trees provide good nesting cover for songbirds.

**Management:** Hackberry is protected by most landowners who recognize its value. This tree is very susceptible to some herbicides used for brush control. Careful management of livestock and white-tailed deer numbers will help maintain this plant and allow for reproduction. Hackberry trees usually get started under the protection of other shrubs after the seeds are deposited in bird droppings.

Bark—The bark of American elm (and some other *Ulmus* species) has deep furrows.

Leaves—Elm leaves are large, with very well-defined veins and a serrated edge. Photos by George Clendenin, USDA-NRCS.

# American elm *(Ulmus americana)*
## ULMACEAE — ELM FAMILY

**Characteristics:** Native; deciduous.

**Growth Form:** Large tree; height to 50 feet.

**Reproduction:** Seeds.

**General Description:** American elm is a stately, beautiful tree that has been admired for its shape and its shade as well as the numerous practical uses for its wood. Elm trees grow along the edges of rivers and large creeks. Fruits are flat, have a thin wing, and fall from the tree in spring.

**Livestock and Wildlife Value:** Browse is highly preferred by livestock and white-tailed deer but is usually out of reach. Seeds are consumed by turkeys, wood ducks, and many kinds of songbirds, as well as squirrels and other small mammals. Sometimes used as a roost tree by turkeys, but the structure of the branches is not ideal for them.

**Management:** Elm is a desirable riparian tree that should be protected. The root system helps stabilize and protect creek and river banks. Riparian areas should be given preferential treatment to encourage the establishment and reproduction of desirable trees. Light or abbreviated grazing periods will allow elm and other riparian trees to establish.

Flowers—Flower spikes are composed of many small white flowers, which are extremely fragrant. Photos by George Clendenin, USDA-NRCS.

# Whitebrush *(Aloysia gratissima)*
## VERBENACEAE—VERVAIN OR VERBENA FAMILY

**Characteristics:** Native; deciduous.

**Growth Form:** Sprawling, multistemmed shrub; height 6–12 feet.

**Reproduction:** Seeds.

**General Description:** Leaves and new twigs appear in opposite matching pairs. Leaves may be shed during very dry periods and reappear following adequate rainfall. Other common names: common beebrush, palo amarillo, cedrón, poleo, cedrón del monte.

**Livestock and Wildlife Value:** Occasionally browsed by goats and white-tailed deer but not preferred. Thickets of whitebrush create very dense cover for a variety of wildlife. Whitebrush, also known as beebrush, is a popular honey-producing shrub.

**Management:** The palatability and nutritional value of the browse is greatly enhanced after prescribed fire, for a time. Whitebrush occasionally gets thick enough in the region to warrant some control. A pelleted herbicide is effective for whitebrush control. Mechanical methods such as grubbing are sometimes used to reduce or control whitebrush.

**Native American Uses:** The Mexican Kickapoo Indians use the leaves and flowers in a tea (Tull 1987).

**Toxic Agent:** Unknown.

**Toxic Description:** Horses, mules, and donkeys are suspected to have been poisoned by this plant. However, horses that are fed well and given mineral and nutritional supplementation are much less likely to consume whitebrush. See *Toxic Plants of Texas* (Hart et al. 2010) for livestock clinical signs and more information.

**Popular Host Trees for Mistletoe**—In this area these include mostly mesquite and hackberry, some bumelia, and a few other trees. Mesquite trees are often a host plant in this region. This is more apparent in the winter, after mesquite has shed its leaves. Photos by George Clendenin, USDA-NRCS.

# Mistletoe *(Phoradendron tomentosum)*
## VISCACEAE—MISTLETOE FAMILY

**Characteristics:** Native; evergreen.

**Growth Form:** Hemiparasitic evergreen plant that grows on the branches of trees.

**Reproduction:** Seeds.

**General Description:** This plant is a hemiparasite, which means it steals water and nutrients from a host plant but still photosynthesizes with its green leaves. Stems and leaves are evergreen, thick, and brittle. Female plants produce the familiar translucent white berries in winter.

**Livestock and Wildlife Value:** All parts of this plant are poisonous to humans, especially the berries. Readily consumed year-round by livestock and white-tailed deer and provides high-quality browse. Berries are eaten by several kinds of songbirds, including cedar waxwings and bluebirds.

**Management:** Mistletoe infestations will often kill affected limbs but will seldom kill the entire tree. Mistletoe is spread by birds that consume the berries.

**Toxic Agent:** Possibly amines and toxic proteins.

**Toxic Description:** The toxins are present in all parts of the plant but especially the whitish fruits. If these are eaten, acute gastroenteritis and heart failure can result (Diggs, Lipscomb, and O'Kennon 1999). Since this plant is often used for Christmas decorations, it is not recommended to bring into the house with small children present. Cattle usually avoid the plant, but they are reported to have died from browsing on the foliage (Tull 1987).

Leaves—The old adage "leaflets three, let it be" helps identify this plant, which is very closely related to sumacs. Photos by George Clendenin, USDA-NRCS.

Why Do I Itch? The itching, redness, and swelling associated with contact with poison ivy are a result of the body's response to urushiol, a very potent oil. Approximately 85 percent of the population will develop an allergic reaction if exposed to poison ivy, oak, or sumac, according to the American Academy of Dermatology. Poison ivy can cause a mild to severe dermatological allergic reaction in some people who touch any part of the plant (direct contact) or touch something that has come into contact with the plant, such as their pants, backpack, or even their pet (indirect contact). They can even have an internal poison ivy reaction by breathing vapors from a burning pile of leaves that contains poison ivy (airborne contact).

**Common Myths:** It is not possible to get this rash from touching someone who has the rash. The skin absorbs the oil too quickly. You cannot get a rash from getting the fluid in the blisters on your skin. Instead, additional spread of the dermatitis comes from unwashed skin, clothing, or other objects, or it can be just a delayed reaction to exposure (American Academy of Dermatology).

Prevention is the best way to avoid a reaction. Wash your hands with soap and cold water as soon as you think you have been in contact. The sooner you wash, the less chance the oil has to bond with the skin. Also wash objects that may have come in contact. If you do become infected or have a severe reaction, seek medical attention.

# VINES

## Poison ivy *(Toxicodendron radicans)*
### ANACARDIACEAE — SUMAC FAMILY

**Characteristics:** Native; perennial.

**Flower Bloom Period:** Early summer.

**Growth Form:** Varies considerably from a low ground cover of root sprouts, to small shrubs, to long vines growing up into the canopies of tall trees.

**Reproduction:** Seeds and root sprouts.

**General Description:** Usually found in bottomlands near creeks and rivers. The leaf is composed of 3 leaflets, each with several lobes or coarse teeth. The leaf shape is extremely variable, which leads some people to incorrectly distinguish some plants as "poison oak" and others as "poison ivy," although botanically they may be the same plant. In addition, the term poison "oak" is a misnomer, as there are no oaks (*Quercus*) that cause a skin reaction. The leaves turn reddish in early fall. The young leaves and stems are also typically reddish. Flowers are inconspicuous; fruits form in fall as loose clusters of small white berries that persist through winter.

**Livestock and Wildlife Value:** Browsed by livestock and white-tailed deer. Berries are eaten by many species of birds.

**Management:** This plant may be targeted for control where it grows near human activity. Otherwise it is a valuable plant for livestock and wildlife. Plant is easily controlled by spot application of herbicide.

**Growth Form—** This wavy-leaf milkvine was found growing intertwined in an algerita plant and would be easy for a casual observer to miss. Photos by George Clendenin, USDA-NRCS.

# Wavy-leaf milkvine *(Funastrum crispum)*
*(Sy = Sarcostemma crispum)*
## ASCLEPIADACEAE—MILKWEED FAMILY

**Characteristics:** Native; perennial.

**Flower Bloom Period:** Spring–early summer.

**Growth Form:** Trailing vine to about 6 feet long, either growing across the ground or more commonly supported by other shrubs.

**Reproduction:** Seeds.

**General Description:** Triangular leaves occur in matching pairs along the stem. Leaves and stems have milky sap. Long, narrow, triangular leaves are greenish gray and have a distinctive wavy edge. Purple flowers occur in hanging clusters. Long, narrow seedpods break apart at maturity, releasing many seeds, each with a tuft of silky hairs. Other milkvines in the Concho Valley include climbing milkweed (*Funastrum cynanchoides*), two-flower milkvine (*Matelea biflora*), and net-vein milkvine (*Matelea reticulata*) (Amos 1998). Other common names: milkvine, wavy-leaf twinevine.

**Livestock and Wildlife Value:** Use by livestock and white-tailed deer is unknown, but the plant is presumed to be of limited value. Silky fluff from the seeds is sometimes gathered by birds for nest lining. This plant is a larval host plant for the queen butterfly. The females lay their eggs only on this and other plants in the milkweed family. The eggs hatch into caterpillars, which feed exclusively on the leaves of milkweed and milkvine.

**Management:** No specific management is identified for this plant.

Photos by George Clendenin, USDA-NRCS.

# Field bindweed *(Convolvulus arvensis)*
## CONVOLVULACEAE — MORNING GLORY FAMILY

**Characteristics:** Introduced; perennial.

**Flower Bloom Period:** Spring–summer.

**Growth Form:** Aggressive vine, forms dense beds.

**Reproduction:** Seeds and extensive spreading rhizomes.

**General Description:** Field bindweed is a problem for the farmer but has also escaped to become a problem for pastureland and home landscapes as well. It is difficult to eradicate because of its fast-growing, aggressive twining nature and its deep roots. It is one of two Texas state-listed noxious plants in this manual (the other being salt cedar). It is also on the state noxious plant lists of Arkansas, Colorado, Kansas, and New Mexico. See the section on noxious and invasive plants in chapter 1 for further information. Leaves are variable in shape but are often triangular with shallow rounded lobes at the base. Flowers are white or pink. Seeds are formed in round dry capsules.

**Livestock and Wildlife Value:** This plant is not common on native rangeland, but white-tailed deer and sheep grazing in cropland fields will consume its leaves.

**Management:** This aggressive and invasive weed from Eurasia can be a serious pest in cropland fields, where it can be difficult to control. Plowing merely spreads the roots and allows new plants to sprout. Do not wait until this one gets away from you; it is easier to eradicate when small than it is after it intertwines among desirable native species. Broadleaf herbicides are used to control this plant.

Flowers—Flowers have 5 points and range from white to pink or white with a red eye in the center. Photos by George Clendenin, USDA-NRCS.

# Texas bindweed *(Convolvulus equitans)*
## CONVOLVULACEAE—MORNING GLORY FAMILY

**Characteristics:** Native; perennial.

**Flower Bloom Period:** Spring–fall in response to rainfall.

**Growth Form:** Forb with trailing stems from a central taproot.

**Reproduction:** Seeds.

**General Description:** Texas bindweed is not to be confused with the aggressive, invasive, introduced field bindweed (*C. arvensis*). The leaves of Texas bindweed are variable in shape and size but are usually long and narrow, with or without lobes or teeth. Plant is covered by fine gray hairs. Seedpods are dry capsules containing up to 4 large, angular seeds.

**Livestock and Wildlife Value:** Grazed by sheep, goats, and white-tailed deer. Seeds are occasionally eaten by quail.

**Management:** Proper grazing management will help maintain the abundance of this plant.

**Flowers—** The flowers of cotton morning glory are pleated or appear to have folds.

**Leaves—** Leaf shape is extremely variable on many of the vines in our region. Here the leaves of cotton morning glory are trilobed. Photos by George Clendenin, USDA-NRCS.

# Cotton morning glory *(Ipomoea cordatotriloba)*
CONVOLVULACEAE—MORNING GLORY FAMILY

**Characteristics:** Native; perennial.

**Flower Bloom Period:** Primarily spring.

**Growth Form:** Aggressive climbing vine, grows on fences or shrubs for support.

**Reproduction:** Seeds.

**General Description:** Many of our vines are hard to identify because the leaf shapes can be extremely variable. Cotton morning glory is no exception: the leaves often have 3 lobes but are sometimes heart shaped. Flowers are numerous, about 2 inches in diameter and pink, lavender, or rose purple. Round seed capsules contain several large, angular seeds. Plant is often seen on fence lines and in gardens and disturbed areas, and sometimes in cropland fields. Other common name: tie vine.

**Livestock and Wildlife Value:** Leaves and stems browsed by livestock and white-tailed deer. Seeds may be eaten by birds.

**Management:** This plant can be aggressive and grow where it is not wanted, including cropland fields. In this case, it can be controlled by herbicides or grubbing of small plants.

Flowers—The large blooms do actually open up to further reveal their beauty; I was just not able to catch one in a photograph!

Vines—Lindheimer morning glory does not have tendrils; rather the stems wrap around objects for support. Photos by George Clendenin, USDA-NRCS.

# Lindheimer morning glory *(Ipomoea lindheimeri)*
## CONVOLVULACEAE—MORNING GLORY FAMILY

**Characteristics:** Native; perennial.

**Flower Bloom Period:** Primarily late spring–early summer.

**Growth Form:** Long, twining perennial vine from a large central taproot.

**Reproduction:** Seeds.

**General Description:** Leaves are about 3 inches across and have 5 deep lobes. Flowers are large and beautiful and about 3 inches in diameter. Like the flowers of many other morning glories, the flowers of this species open in the morning and begin to close by noon. They are soft blue, turning lavender. Stems may climb 8 feet up into shrubs, which are used as support. Other common name: blue morning glory.

**Livestock and Wildlife Value:** Leaves and stems are browsed by livestock and white-tailed deer. Seeds may be eaten by birds.

**Management:** This plant requires no specific management. Rootstock can be easily transplanted by those who wish to add this attractive vine to their landscape.

Photos by George Clendenin, USDA-NRCS.

Female flower.

Male flower.

# Buffalo gourd *(Cucurbita foetidissima)*
CUCURBITACEAE — GOURD FAMILY

**Characteristics:** Native; perennial.

**Flower Bloom Period:** Spring–summer.

**Growth Form:** Large, coarse vine, trails across the ground or grows on fences. Plants are often 20 feet or more in diameter.

**Reproduction:** Seeds.

**General Description:** Leaves are very large, blue green, and triangular, with toothed edges. Leaves and stems have a rough, scratchy texture. The entire plant has a strong disagreeable odor. Flowers are large and yellow. Fruits are hard round gourds about the size of a tennis ball, turning yellow at maturity. Gourds contain many white seeds shaped like watermelon seeds. Buffalo gourd has a massive taproot. Other common names: stinking gourd, calabacilla loca, wild gourd.

**Livestock and Wildlife Value:** Although the plant has a strong odor and rough texture, it is grazed by livestock. White-tailed deer consume the young tender leaves. As the gourds deteriorate or break apart, the seeds are made available and are eaten by some species of birds.

**Management:** No specific management is identified for this plant.

**Native American and Other Uses:** According to archaeological evidence, Native Americans have used buffalo gourd for over 9,000 years (Tull 1987). Various Indian tribes throughout the Southwest and Mexico have used the dried and roasted seeds for food, the oil as a cosmetic, and the green fruits and roots as a laundry detergent. The roots have a high content of saponin, which yields a cleansing soapy lather. In addition, the dried fruits (gourds) have been used in ritualistic gourd dances (Tull 1987). The Dakota and Omaha-Ponca tribes also used the roots medicinally, referring to them as "pumpkin medicine" or "human being medicine," respectively (Gilmore 1919).

Fruit—These soft, fleshy gourds are about 1 inch in diameter. Gourds are green and striped when young, resembling a small watermelon, and turn bright red at maturity. Photos by George Clendenin, USDA-NRCS.

# Balsam gourd *(Ibervillea lindheimeri)*
## CUCURBITACEAE — GOURD FAMILY

**Characteristics:** Native; perennial.

**Flower Bloom Period:** Spring–summer.

**Growth Form:** Climbing vine with tendrils from a large taproot, usually grows with the support of fences, brush piles, shrubs, or trees.

**Reproduction:** Seeds.

**General Description:** Leaves are thick and succulent and variable in shape, with 3 or 5 lobes. Flowers are small and yellow. Each gourd contains several seeds within the slimy yellow pulp. A close relative of this vine that is also found in the region is slim lobe balsam gourd (*Ibervillea tenuisecta*), which has narrower leaves and much smaller fruits. Other common names: snake apple, globe berry.

**Livestock and Wildlife Value:** Value to livestock and white-tailed deer is unknown.

**Management:** No specific management is recommended for this plant.

The large single seed inside each berry resembles a snail.
Photos by George Clendenin, USDA-NRCS.

# Carolina snailseed (Cocculus carolinus)
## MENISPERMACEAE — MOONSEED FAMILY

**Characteristics:** Native; perennial.

**Flower Bloom Period:** Summer.

**Growth Form:** Long vine from a woody base, grows 6–12 feet up into trees and bushes.

**Reproduction:** Seeds and root sprouts.

**General Description:** Leaves are variously shaped, usually rounded triangles with or without deep or shallow lobes, and alternately arranged along the stem. The undersides have soft, fine hairs. Small white or cream flowers form in spikes in summer. Berries begin to ripen in late summer, when they turn a brilliant red. Plant is sometimes confused with greenbriar; both have similarly shaped leaves, but snailseed is entirely free of thorns and does not have tendrils. Although little documentation exists, the berries are presumed toxic to humans (Tull 1987).

**Livestock and Wildlife Value:** Readily browsed by livestock and white-tailed deer. Berries, which are produced in profusion, remain intact on the vines for several months and are eaten by a large variety of songbirds.

**Management:** Carolina snailseed benefits from good conservative grazing management and control of white-tailed deer numbers. Under good management, vines will increase as a result of prolific root sprouting. This plant almost always grows up out of the protection of other shrubs.

Flowers—The feathery flowers are the source of the common names "goat beard" and "old man's beard."

This old man's beard persists on this fence line only because goats are absent in this pasture! Photos by George Clendenin, USDA-NRCS.

# Old man's beard *(Clematis drummondii)*
## RANUNCULACEAE—BUTTERCUP FAMILY

**Characteristics:** Native; perennial.

**Flower Bloom Period:** Late spring–early summer.

**Growth Form:** Straggly, viny shrub from a woody base, usually growing up into other shrubs and trees or fences for support.

**Reproduction:** Seeds.

**General Description:** Leaves are in opposite matching pairs along the stem and are composed of 3 smaller leaflets, each with several pointed lobes. Small yellowish-white flowers form in clusters in spring. Seeds are formed in tight clusters. Each seed is attached to a long feathery plume. As the seeds mature, they are carried away by the wind. Other common names: goat beard, barbas de chivato.

**Livestock and Wildlife Value:** Readily browsed by livestock and white-tailed deer. Dense tangles of stems and vines provide good protective cover for small birds.

**Management:** Old man's beard and other desirable browse plants can be favored by conservative grazing management and careful attention to livestock and white-tailed deer numbers. Since this vine is often confined to the cover of other shrubs, it is somewhat protected from heavy browsing. Under good management, this plant can be expected to spread and increase away from protective cover.

**Native American and Other Uses:** The vines have been used to make baskets, although *Clematis* vines are known to contain highly caustic, acrid juices; therefore care should be used when handling. As a source of dye, the leaves and vines of *C. drummondii* yield outstanding colors of golden brown, brown, and yellow (Tull 1987). The feathery plumes were used to help start campfires, and some Native Americans used them as swaddling for babies. The plant was used medicinally as a tea for migraine headaches (Moore 2003).

Community—Greenbriar is sometimes confused with Carolina snailseed; both have similarly shaped leaves, but greenbriar is heavily armed with thorns and has tendrils, while Carolina snailseed has neither thorns nor tendrils. Photo by George Clendenin, USDA-NRCS.

# Greenbriar *(Smilax bona-nox)*
## SMILACACEAE — GREENBRIAR FAMILY

**Characteristics:** Native; perennial.

**Flower Bloom Period:** Spring.

**Growth Form:** Large, tough, spiny vine, grows up to 20 feet into trees, although sometimes so heavily browsed that growth form is a small, spindly shrub.

**Reproduction:** Seeds and root sprouts.

**General Description:** If you run across this plant while walking through wooded areas or river bottoms, you will remember it. Stems are armed with short, stout thorns. New stem tips produce tendrils that wrap around nearby objects for support. Leaves are generally triangular, with rounded lobes and rounded or pointed tips, and are alternately arranged along the stem. Greenbriar is deciduous, although leaves often persist into early winter. Root system is composed of very large knotty rhizomes. The blue-black berries of *S. bona-nox* are not considered toxic to humans, but they are also not considered edible (Tull 1987).

**Livestock and Wildlife Value:** Extremely good browse for livestock and white-tailed deer. Berries are eaten by a number of kinds of songbirds.

**Management:** Greenbriar can be maintained with good production by proper management of goat and white-tailed deer numbers. Heavy hedging of this plant often indicates too many goats or white-tailed deer. This plant responds favorably to prescribed burning with large amounts of new growth. However, if supporting shrubs and trees are burned, greenbriar becomes vulnerable to overbrowsing.

Photos by George Clendenin, USDA-NRCS.

# Ivy treebine *(Cissus incisa)*
## VITACEAE—GRAPE FAMILY

**Characteristics:** Native; perennial.

**Flower Bloom Period:** Summer.

**Growth Form:** Large climbing vine, grows up into trees, shrubs, and brush piles for support.

**Reproduction:** Seeds.

**General Description:** Thick, succulent leaves are deeply divided into 3 segments, often folded; leaf edges are lobed or serrated. Leaves sometimes have unpleasant odor. Opposite from each leaf is a tendril that grasps and curls around objects for support. Flowers occur in large clusters in summer. Fruits are clusters of blue or black berries that ripen in late summer. Other common names: possum grape, cow itch.

**Livestock and Wildlife Value:** Occasionally browsed by white-tailed deer and goats but not preferred. Berries are eaten by many species of birds.

**Management:** This plant requires no specific management but is more likely to occur where there are other shrubs or brush piles for support.

Photos by George Clendenin, USDA-NRCS.

# Virginia creeper *(Parthenocissus quinquefolia)*
## VITACEAE — GRAPE FAMILY

**Characteristics:** Native; perennial.

**Flower Bloom Period:** Spring.

**Growth Form:** Climbing vine, grows with the support of shrubs and trees.

**Reproduction:** Seeds.

**General Description:** This plant is often confused with poison ivy and may grow alongside poison ivy on the same tree. However, unlike poison ivy it does not cause contact dermatitis. The berries, however, are suspected to be extremely poisonous, especially to children, because of high concentrations of oxalic acid, and the plant tissues have microscopic crystals that may irritate the skin (Diggs, Lipscomb, and O'Kennon 1999; Tull 1987). Leaves are composed of 5 leaflets (sometimes 7) radiating out from a central point. Each leaflet is serrated on the edge. Leaves turn brilliant orange or red in fall. Small green flowers form in spring. Clusters of round blue berries ripen in late summer. This plant is also planted as a landscape vine because of its attractive foliage.

**Livestock and Wildlife Value:** Readily browsed by livestock and white-tailed deer. Berries are eaten by many kinds of birds.

**Management:** This vine can be maintained or increased by good conservative grazing management, proper control of white-tailed deer numbers, and adequate woody cover for it to grow on.

Photos by
George Clendenin,
USDA-NRCS.

# CACTI AND AGAVES

## Sacahuista *(Nolina texana)*
### AGAVACEAE — CENTURY-PLANT FAMILY

**Characteristics:** Native; perennial; evergreen.

**Growth Form:** Large, robust, clumpy plant, resembling a large bunchgrass; diameter to 8 feet and height to 2 feet.

**Reproduction:** Seeds.

**General Description:** Plant usually grows on very shallow or rocky soil and hillsides. Many long, tough, narrow leaves emerge from a very large basal stump. Although plants resemble a large clump of coarse grass, they are actually more closely related to yucca. Leaves arch over and are more horizontal than vertical. When moisture is above average, flower stalks with hundreds of tiny white to pinkish flowers form in spring. Flower stalk does not elongate as in yucca but stays below the top of the plant, partially hidden by the leaves. Many round white seeds are formed within the seedhead. Other common name: beargrass.

**Livestock and Wildlife Value:** Flowers are sometimes eaten by livestock, especially if other green forage is scarce. Toxic plant; see toxicity information. Cattle and deer sometimes eat the coarse leaves, especially the fresh new growth after a fire. Large clumps can provide nest concealment for turkeys. Many ranchers say that the only thing sacahuista is good for is to hide rattlesnakes.

**Management:** Sheep and goat raisers should note when sacahuista is beginning to bloom and remove stock from those areas until blooming is complete. Cattle can be moved into pastures that have heavy blooms, because they are more resistant. Burning followed by cattle grazing can temporarily reduce the coverage of sacahuista. Plants sometimes die after this treatment. Plants can be controlled by application of pelleted herbicide or mechanical grubbing.

**Toxic Agent:** Saponins.

**Toxic Description:** Only the flower buds, blooms, and fruits are toxic; the green foliage appears not to be. Ingesting these buds, blooms, and fruits causes severe liver damage to livestock. Sheep and goats are more susceptible to poisonings than cattle are. Deer are unaffected. See *Toxic Plants of Texas* (Hart et al. 2010) for livestock clinical signs and more information.

Buckley yucca (*Yucca constricta*) has many slender leaves, usually ¼ to ½ inch wide, and threads coming from the leaves.

# Buckley yucca *(Yucca constricta)*
AGAVACEAE—CENTURY-PLANT FAMILY

**Characteristics:** Native; perennial; evergreen.

**Growth Form:** Clumpy plants with coarse, spine-tipped leaves; flower stalk height to 8 feet.

**Reproduction:** Seeds and rhizomes.

**General Description:** Two common yucca species occur in the region, but for practical and management purposes they can be considered together. Buckley yucca (*Yucca constricta*) has many slender leaves, usually ¼ to ½ inch wide, and threads coming from the leaves. Plateau yucca (*Y. reverchonii*) has fewer but wider leaves to about 1 inch wide, has no threads or hairs coming from the leaves, and is more cluster forming. It is observed to be less prevalent in the Concho Valley than Buckley yucca. In both species, flower stalks begin to form from the center of the plant in spring. Numerous large white flowers appear as stalks mature. Seedpods form in summer, turn brown and woody, and contain many flat black seeds. Yucca forms very massive root systems, including rhizomes that form new plants. Amos (1998) cites a third yucca in her inventory of the Concho Valley, *Y. torreyi*.

**Livestock and Wildlife Value:** Young tender flower stalks are relished by all classes of livestock and deer. Nutritional value of soft growing stalks is very high. Quail sometimes use the edges of yucca plants for nest concealment. Horses have been known to dig up yucca and eat the roots. Interestingly, yucca is incorporated into many hoof care and joint supplements because it is thought to have anti-inflammatory properties. There is no published information on clinical trials at this time that would support or deny these assertions.

**Management:** If yucca gets too abundant, it can take up too much space and reduce forage production and forage availability, but this is not a common problem. At moderate densities, the food value of yucca stalks probably compensates for any loss of grass. Yucca increases in density after fire because of the stimulation of rhizome production.

Plateau yucca (*Yucca reverchonii*) has fewer but wider leaves to about 1 inch wide, has no threads or hairs coming from the leaves, and is more cluster forming. Photos by George Clendenin, USDA-NRCS.

**Native American Uses:** There is a long list of uses of yucca by native peoples of many tribes. It has been said that yucca ranked foremost among the wild plants used by peoples of the Southwest (Diggs, Lipscomb, and O'Kennon 1999). Yucca was used for soap, shampoo, food, medicine, and ceremonial purposes. It was also used for sewing needles as well as thread. It is still used by New World peoples in the cosmetic and soap industry, and as a home remedy tea for arthritis (Moore 2003; Diggs, Lipscomb, and O'Kennon 1999).

Fruits are bright red at maturity, 1 inch in diameter, and full of large black seeds.

There is a reason for the common names "horse crippler" and "devil's pincushion"; the rigid spine clusters can injure grazing animals, and the plant is low to the ground and hard to see. Photos by George Clendenin, USDA-NRCS.

# Horse crippler *(Echinocactus texensis)*
CACTACEAE—CACTUS FAMILY

**Characteristics:** Native; perennial.

**Growth Form:** Large, low cactus; diameter to 12 inches and height 1–4 inches above the ground.

**Reproduction:** Seeds.

**General Description:** Hard-bodied cactus with well-defined ridges and large, stout spines. Multiple flowers emerge from the center. Flowers are very large and showy, with pink petals and a maroon center. Plants may rise a few inches above the surface or be nearly flat, with the main body of the cactus underground. Other common names: devil's pincushion, devil's head, manca caballo.

**Livestock and Wildlife Value:** Fruits and/or seeds are eaten by some kinds of birds and small mammals.

**Management:** This and other cacti are susceptible to herbicides used to control prickly pear. Mechanical disturbance and damage from machinery is likely to cause injury or death.

**Flowers**—The very large, beautiful, purple-violet flowers form in spring.

**Growth Form**—Strawberry cactus is clump forming. Photos by George Clendenin, USDA-NRCS.

# Strawberry cactus *(Echinocereus enneacanthus)*
## CACTACEAE — CACTUS FAMILY

**Characteristics:** Native; perennial.

**Growth Form:** Low, multistemmed, clump-forming cactus; height 6–12 inches, diameter to 2 feet.

**Reproduction:** Seeds.

**General Description:** Plants have few to many upright cylindrical stems with vertical ribs and abundant long spines. Fruits are round, about 1 inch in diameter, and covered by short spines. As the fruits ripen in summer, the spines fall off and the fruits turn pinkish. Once the spines are completely removed, the fruits are sweet, juicy, and very good to eat, having a strawberry-like taste. Plant grows mostly in shallow, rocky soil. Other common name: pitaya.

**Livestock and Wildlife Value:** Fruits are eaten by birds and small mammals, although the matrix of spines limits availability.

**Management:** This and other cacti are susceptible to herbicides used to control prickly pear. Mechanical disturbance and damage from machinery is likely to cause injury or death. This attractive cactus can be propagated for ornamental use. Large clumps can be dug and divided, or individual stems can be cut, allowing the cut to completely dry up before rooting in soil.

Flowers—These pink to light purple flowers are very beautiful.

Spines—This cactus can easily be handled without feeling any sharp spines. Photos by George Clendenin, USDA-NRCS.

# Lacy cactus *(Echinocereus reichenbachii)*
CACTACEAE — CACTUS FAMILY

**Characteristics:** Native; perennial.

**Growth Form:** Low, "barrel type" cactus; height 3–6 inches.

**Reproduction:** Seeds.

**General Description:** Each stem has numerous vertical ridges or ribs. Ribs are covered by a network of short spines that lie flat against the surface of the cactus. Flowers are large and beautiful and form in spring. Fruits are round and covered by short spines. Other common name: lace cactus.

**Livestock and Wildlife Value:** Fruits are eaten by a variety of birds and small mammals.

**Management:** This and other cacti are susceptible to herbicides used to control prickly pear. Mechanical disturbance and damage from machinery is likely to cause injury or death.

Photos by George Clendenin, USDA-NRCS.

# Prickly pear *(Opuntia engelmannii)*
## CACTACEAE—CACTUS FAMILY

**Characteristics:** Native; perennial.

**Growth Form:** Large cactus with multiple flat pads; height 1–5 feet, diameter to 10 feet.

**Reproduction:** Seeds and rooting pads.

**General Description:** This ubiquitous and familiar cactus needs no description to the residents of the region. There are several different species of prickly pear in the region, which are often lumped together for management purposes. Size and color of pads, spines, flowers, and fruits vary considerably. The larger upright types are generally referred to as Engelmann prickly pear or Lindheimer prickly pear. The low, prostrate types, sometimes called "running pear," are different species, either brown spine prickly pear or plains prickly pear. Precise identification is confused by the frequent hybridization between species.

**Livestock and Wildlife Value:** Prickly pear is valuable for wildlife and livestock. It may be considered overabundant, yet it is underappreciated. The pads are frequently eaten by livestock and deer, especially in periods of stress when other foods are in short supply. The pads are high in moisture and have good energy content but poor protein content. The sweet, juicy, purple or red fruits, commonly called pear apples or tunas, are readily eaten in summer by all classes of livestock, deer, rabbits, and other small mammals. Turkeys sometimes eat an entire fruit or break it apart to eat in pieces. Quail and other birds sometimes peck into a fruit to get seeds and/or moisture. Clumps of prickly pear provide alternate nesting cover for quail and can be important when grass cover is inadequate or grazed short. Prickly pear also provides physical protection and refuge areas for desirable grasses, forbs, and shrubs.

**Management:** Despite its considerable value, prickly pear has become a serious problem on many ranches where it has increased to excessive and undesirable densities. Excessive densities of this plant render large areas inaccessible to grazing animals and materially reduce the availability of forage. Furthermore, the consumption of pads or fruits sometimes causes injury to the tongue, mouth, or digestive tract of livestock, especially sheep and goats, which can affect their health and productivity.

In addition to providing occasional forage for livestock, prickly pear also provides physical protection and refuge areas for desirable grasses, forbs, and shrubs. Photo by Dee Ann Littlefield, USDA-NRCS.

While prescribed burning can be an effective method for managing prickly pear, a single fire does not usually kill all the plants. A series of fires can be very effective. Photo by Dee Ann Littlefield, USDA-NRCS.

Prickly pear infestations may be managed in several ways. The most common method of control is with herbicides, applied either aerially or as individual plant treatments. The primary herbicide that has been used for prickly pear control remains active in the soil for several years and is harmful to many other species of forbs and woody plants, especially hackberry, littleleaf sumac, and wolfberry. Individual spot treatment reduces the nontarget damage but is feasible only with light infestations. Prescribed burning can also be very effective in reducing infestations, with hotter fires more damaging than cooler fires. A single fire seldom provides satisfactory control, but a series of several fires can provide a good level of control. Sometimes, lower rates of herbicides are used in conjunction with fire. Ranchers will sometimes turn livestock into a pasture immediately following a fire to consume freshly burned pads. This provides a twofold benefit—a reduction in pads and a short-term source of forage. Mechanical methods of control are sometimes used and usually involve raking pads into large piles.

Many natural enemies of prickly pear periodically cause reductions or die-offs. Several insects bore into or puncture pads, sometimes causing secondary infections and rot. Wet conditions often promote various fungal diseases that can eliminate entire plants in a short period.

Prickly pear is spread in 3 important ways. The most obvious is the scattering of the pads that inevitably takes place during mechanical brush-control operations. As plants are disturbed, pads are scattered across the ground, many of them eventually rooting and forming new plants. I have witnessed that if prickly pear is present and disturbed during brush removal of mesquite or juniper, in 3 years the prickly pear problem will come back worse than before. Another, less obvious method is the spreading of pads and pieces of pads by livestock and deer as they consume them. Pieces of pads as small as 1 inch in diameter will sprout roots and grow into new plants. The third method involves the spreading of seeds eaten by livestock and wildlife. The seeds inside the fruit are not generally digested but pass through the animal intact. Examination of cow pies and the pellets of goats, sheep, deer, and rabbits will confirm that seeds are spread across pastures in large quantities as a result of these animals consuming the fruits.

Photos by George Clendenin, USDA-NRCS.

# Tasajillo *(Opuntia leptocaulis)*
## CACTACEAE—CACTUS FAMILY

**Characteristics:** Native; perennial.

**Growth Form:** Upright, much-branched cactus shrub; height to 4 feet.

**Reproduction:** Seeds and rooting joints.

**General Description:** Plants are composed of hundreds if not thousands of joints, each about the diameter of a pencil and varying in length from ½ inch to 6 inches. Spines can be very short to very long. The long spines often have an external sheath. Flowers are small and yellow, forming in spring. Fruits are small and bright red, with small patches of tiny spines. Fruits remain on plants for many months, often into winter. Other common names: turkey pear, Christmas cactus.

**Livestock and Wildlife Value:** Fruits are readily eaten by turkeys, quail, and other birds. Joints are sometimes eaten by sheep, goats, deer, quail, and turkeys, especially when other green forage is scarce.

**Management:** Plant often becomes overabundant and can form dense thickets that inhibit movement and access by livestock and humans. Tasajillo is very susceptible to fire, which is the most cost-effective method of control. Tasajillo can also be controlled with herbicides. It spreads easily by the scattering of joints, which can root and form new plants.

# Comparing and Contrasting Similar Plants

This section is dedicated to identifying similar plants in the region, with the help of side-by-side photographic comparison. I have included a few of the plants that are most often confused or hard to distinguish in the Concho Valley with a brief description of their similarities and differences.

# Comparing Western Ragweed (Ambrosia psilostachya) and Field Ragweed (Ambrosia confertiflora)

**Western ragweed (*Ambrosia psilostachya*)**—Has much smaller and less intricately divided leaves. Taxonomists call this leaf shape pinnatifid because it resembles a pinnately compound leaf. This plant is far less common than field ragweed in the Concho Valley region. Photo by George Clendenin, USDA-NRCS.

**Field ragweed (*Ambrosia confertiflora*)**—This is still considered one leaf; however, leaves can be much larger and divided even more intricately. Taxonomists still call this leaf shape pinnatifid, even though it is parted many times within the same leaf. This plant is far more common in the region than western ragweed. Photo by George Clendenin, USDA-NRCS.

# Comparing Bush Sunflower *(Simsia calva)* and Orange Zexmenia *(Wedelia texana)*

**Bush sunflower** (*Simsia calva*) — The flower is more yellow than orange and has more ray flowers (petals) than that of orange zexmenia. Leaves are broadly triangular, and unlike those of orange zexmenia, the leaves of bush sunflower are sometimes perfoliate, which means that opposite leaves appear to be attached or growing together at the base (see left picture). Photos by George Clendenin, USDA-NRCS.

**Orange zexmenia** (*Wedelia texana*) — The flower is more orange than yellow and has fewer ray flowers (petals) than that of bush sunflower. Leaves are narrowly triangular as compared to the broader triangular leaves of bush sunflower. Photos by George Clendenin, USDA-NRCS.

# Comparing the Catclaws: Acacia versus Mimosa

**Catclaw acacia** (*Acacia greggii*)
Shape: Large shrub to small tree. Often nice canopy.
Leaves: Bipinnate, 1–3 pairs of pinnae, 3–7 pairs of leaflets per pinna.
Flowers: Elongated and cream colored.
Thorns: Recurved.
Beans: Large and twisted.
General: Usually the most common of the 4 in this region.
Photos by George Clendenin, USDA-NRCS.

**Catclaw mimosa** (*Mimosa aculeaticarpa* var. *biuncifera*)
Shape: Often nondescript small shrub.
Leaves: Bipinnate, 4–8 pairs of pinnae per leaf, 5–12 pair of leaflets per pinna. Very small leaflets and pinnae.
Flowers: Round, sparse, white puffballs. Sometimes tinged with pink.
Thorns: Nastiest thorns! Usually in pairs.
Beans: Small, often with recurved prickles on the margins.
General: The nastiest thorns of the 4 species shown here!
Photo by Steve Nelle, USDA-NRCS.

**Roemer's acacia** (*Acacia roemeriana*)
Shape: Small shrub to small tree.
Leaves: Bipinnate, 1–3 pairs of pinnae, 4–8 pairs of leaflets per pinna. Largest of the 4 species shown here.
Flowers: Round, cream-colored puffballs.
Thorns: Recurved.
Beans: Large, copper colored.
General: Best browse of the 4 catclaws. Twigs sometimes reddish.
Photos by George Clendenin, USDA-NRCS.

**Fragrant mimosa** (*Mimosa borealis*)
Shape: Often nondescript small shrub.
Leaves: Bipinnate, 1–4 pairs of pinnae per leaf, 3–8 pairs of leaflets per pinna.
Flowers: Round, rose to pink, with yellow tips.
Thorns: All along the stem.
Beans: Often without recurved prickles on the margin; sometimes with an occasional prickle.
General: Most attractive flower!
Photos by George Clendenin, USDA-NRCS.

# Comparing Western Soapberry *(Sapindus)*, Pecan *(Carya)*, Little Walnut *(Juglans)*, and Flameleaf Sumac *(Rhus)*

**Western soapberry (*Sapindus saponaria* var. *drummondii*)**
This is one leaf comprised of many leaflets (pinnately compound). There are an even number of leaflets (common in *Sapindus*), and no terminal (end) leaflet (most of the time). Leaf edges are smooth (or nearly so), unlike those of *Carya* and *Juglans,* which are slightly to strongly serrated.
Photo by George Clendenin, USDA-NRCS.

**Pecan (*Carya illinoinensis*)**
This is one leaf comprised of many leaflets (pinnately compound). The leaf edges are strongly serrated, and there is usually a terminal (end) leaflet. The terminal leaflet does not have to be smaller than the middle lateral leaflets, as in *Juglans*. Lateral leaflets are usually sickle shaped.
Photo by George Clendenin, USDA-NRCS.

**Little walnut (*Juglans microcarpa*)**
This is one leaf comprised of many leaflets (pinnately compound). All leaves in *Juglans* are alternate and pinnately compound. The mature leaflets of little walnut are slightly serrated on the edges. The end (terminal) leaflet in *Juglans* is always smaller than the middle lateral leaflets.
Photo by George Clendenin, USDA-NRCS.

**Flameleaf sumac (*Rhus copallinum*)**
This is one leaf comprised of many leaflets (pinnately compound). The most unique feature of this leaf is that the main stem (rachis) has a leafy wing, which almost makes the stem appear flattened. Leaflets are mostly smooth on the edges.
Photo by George Clendenin, USDA-NRCS.

# Section Two

# Rangeland Management in West Central Texas

A well-managed riparian area in the Concho Valley. Photo by Mark Meyer, USDA-NRCS.

# Historical Accounts of Vegetation and Landscape, 1683 to 1858

There is much discussion and debate about what the Concho Valley vegetative landscape looked like historically. Range scientists and ecologists refer to vegetation that was once found in a region as "historic" or "climax" vegetation. Climax vegetation, in theory, is a somewhat "stable" plant community that persists over an extended period and will continue to persist if not affected by some disturbance. These "disturbances" can be natural, environmental, or human-caused. Also associated with this popular theory is the concept that plant communities can change from one state to another based on these disturbances. Plant communities can undergo "succession" or can "transition" to climax communities or other states based on change. Climax vegetation becomes a big focus because management decisions, good or bad, may affect how a community changes. There are some challenges that accompany climax vegetation theory: What time period do we designate as climax? Do we know for certain what really grew in a given place at that time? Does one area where evidence is found represent a region on a larger scale? Why did the vegetation change?

Determining climax vegetation may be difficult and problematic. These historical vegetation or plant lists are assembled by thoroughly evaluating a site's physical and scientific evidence (climate, soils, geology, plant physiology, seed dispersal mechanisms, etc.), considering archaeological evidence such as packrat middens (these mummified "garbage piles" of packrats and other animals sometimes contain abundant fossilized leaves, seeds, and fruits), comparing similar sites that have received varying degrees of use, reviewing early historical and botanical records, and plenty of theorizing. It is not my intent in this chapter to promote, discuss, or defend climax vegetation theory. I merely mention it briefly to introduce the following historical literature and place it within the framework of ecological change.

The following four eyewitness accounts provide a glimpse of the Concho Valley landscape historically. These accounts, dating from 1683 to 1858, were all written before the establishment of Fort Concho in 1867 and before the introduction of domestic livestock. I have chosen to include accounts only up to 1858 because the landscape of the Concho Valley began changing shortly thereafter.

The first account is from Juan Domínguez de Mendoza, a Spanish explorer who recorded his eastbound journey through the Concho

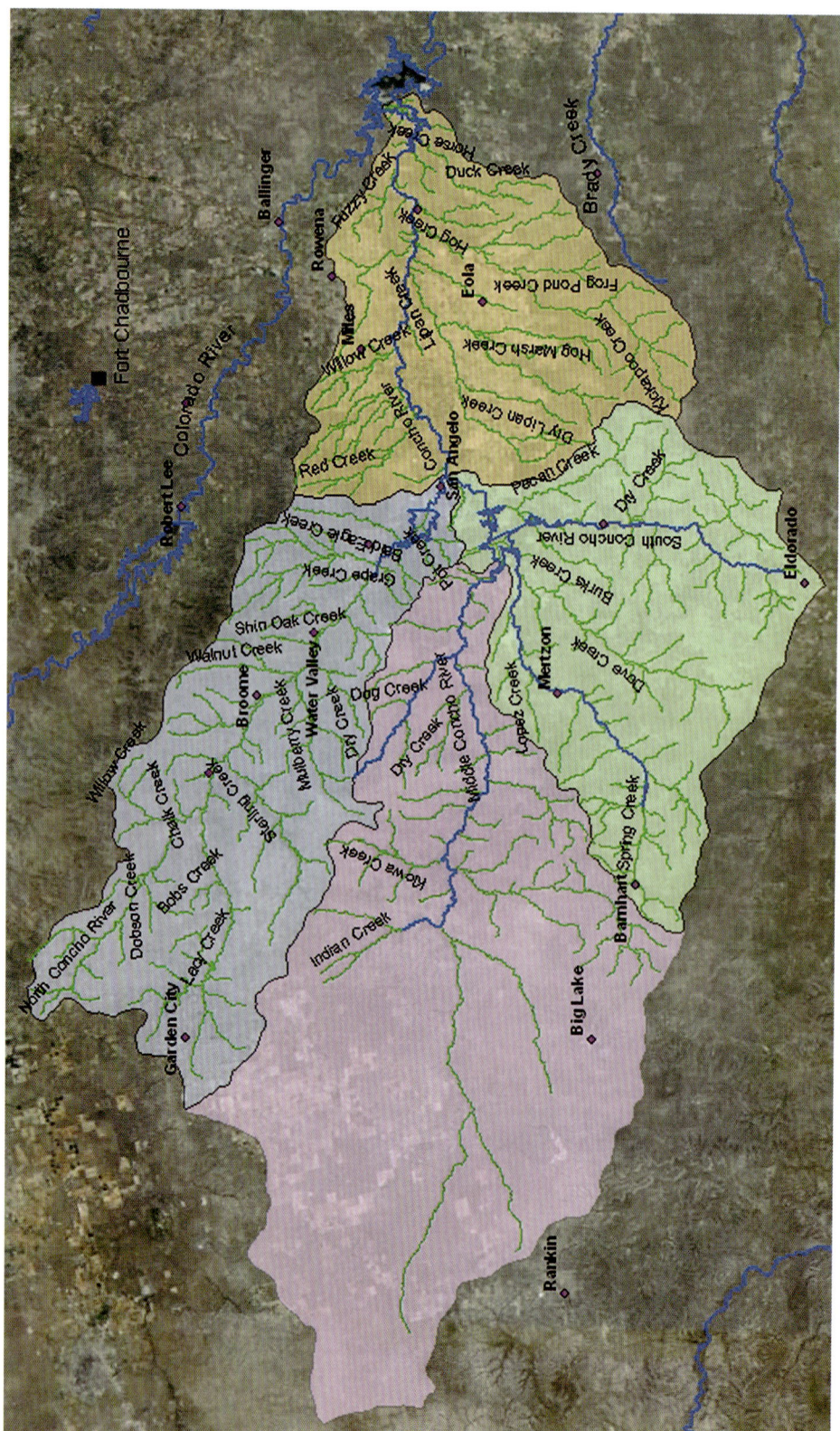

Main rivers and named streams of the Concho Valley. This map can be used as a reference for the historical accounts. Town names are shown for reference but may not have been established at the date of the account. ArcGIS map by George Clendenin.

Valley in 1863. The second account is from Lt. F. T. Bryan of the US Army Topographical Engineers. Lieutenant Bryan was one of several engineers surveying western Texas for possible routes for a transcontinental railroad. He was assigned a route from Fredericksburg through the Concho Valley, then across the Pecos River to El Paso del Norte. The third account is from John R. Bartlett, who led the US-Mexican Boundary Survey team through the Concho Valley area on the way to El Paso (Maxwell 1979). Finally, the fourth account is taken from the writings of Waterman L. Ormsby, a special news correspondent of the *New York Herald*. Ormsby was the only through passenger aboard the Butterfield Overland Mail Stagecoach, the first westbound stagecoach traveling from Saint Louis to San Francisco. His total journey was over 2,700 miles. He wrote a series of eight articles about his journey, which were eventually published in the *Herald* (Ormsby 1942). The accounts follow in chronological order.

**Mendoza**—Records his eastbound journey down the Middle Concho River to the juncture of the three Concho Rivers. This is the earliest known written description of the vegetation of the Concho Valley.

2 February 1683

SITE: Mouth of Kiowa Creek, after camping in Centralia Draw near the present head of the Middle Concho River.

ACCOUNT: *"In this place were the first pecan trees we saw, for its bottoms have many groves of them; many nuts were gathered. . . . It also has shells, a variety of fish, and very lofty live oaks, so large that carts and other bulky things can be made of them. There is a great variety of plants and of wild hens which make noise at dawn. The river bottoms are very extensive and fertile, in its groves are many grape vines and springs, and many prickly pear patches; and all of the foregoing are on both sides of the river."*

11 February 1683

SITE: Site of Sherwood, Texas, the former county seat of Irion County, located about a mile east of Spring Creek.

ACCOUNT: *"The place is in a plaza which has several great groves of very tall pecan and live oak trees. There are a number of wild hens and other kinds of game."*

22 February 1683

SITE: Juncture of the three Concho Rivers, present site of San Angelo.

ACCOUNT: *"On both sides are great bottoms; there is a great luxuri-*

ance of plants, nuts, and other kinds of trees, and wild grapes, good pasturage, a variety of birds, and wild hens."

Mendoza's account makes no mention of mesquite or juniper. He does mention many groves of tall pecan and live oak trees, mature enough to be fruit bearing. He also mentions prickly pear at least being common. The "wild hens" mentioned here and elsewhere most likely refer to turkeys, which prefer the heavily wooded bottoms mentioned in this description. Also notable are the many springs and the extensive and fertile river bottoms.

**Lt. F. T. Bryan, US Army Topographical Engineers**—Records his westbound journey through the Concho Valley.

28 June 1849

SITE: Westbound from the South Fork of Brady's Creek in southern Concho County to the head of Brady's Creek.

ACCOUNT: *"Marched through a beautiful country to the headwaters . . . through a prairie covered with scattered mesquite and mesquite grass. There is abundance of wood for culinary purposes and the grass is abundant and good for grazing."*

Mesquite is first mentioned in this account from 1849. It is also important to note that the grass is still "abundant and good for grazing," most likely for the horses on the journey since sheep and cattle were not introduced until 1877 and 1880, respectively. The primary grazers at this time would have been the buffalo. We do not know for sure what type of grass "mesquite grass" refers to, but it could be any grass that grows well in a mesquite community. Many local ranchers now talk about "mesquite grass," which is actually Texas wintergrass (*Nassella leucotricha*). Texas wintergrass currently often grows in abundance in scattered mesquite communities.

29 June 1849

SITE: Westbound from Brady's Creek to Kickapoo Creek.

ACCOUNT: *"The country between Brady's Creek and Kickapoo is admirably adapted for a natural road. Our whole route was over an open, level, mesquite prairie requiring nothing but traveling to make a road in any direction. The timber is mesquite, only enough for cooking. The grazing was excellent—dog towns and rattlesnakes abounded most of the way."*

The mention of "dog towns" and a later reference to "prairie dogs . . . in abundance" is revealing. Recent writings have suggested the demise of the prairie dog as a reason for the rapid increase of mesquite. Extensive prairie dog towns are known for having a profound effect on vegetation, and prairie dogs could have been responsible for eating the mesquite and other shrub seedlings as soon as they emerged. The rattlesnakes, however, have been preserved and sustained until the present time!

30 June 1849

SITE: Westbound from Kickapoo to Lipan Creek.

ACCOUNT: *"Crossed Kickapoo without any difficulty beyond cutting out the brush. Country becomes hilly, stony, and barren, being a succession of gentle elevations and depressions covered with broken pieces of limestone. . . . In several places the grazing seemed very good but there was an almost total absence of anything like timber. We passed Potato Spring at 2.5 miles from Kickapoo. Timber of pecan and live oak are very heavy on Lipan Creek."*

1 July 1849

SITE: Westbound toward Pecan Creek.

ACCOUNT: *"Country is high, rolling, and stony except in valleys where there is excellent mesquite grass. Timber on the banks of this creek is pretty large. Grazing is good. . . . Rattlesnakes and prairie dogs continue in abundance."*

2 July 1849

SITE: Westbound, crossing the South Concho River and camping at Dove Creek.

ACCOUNT: *"There is an almost total absence of timber. Now and then there is a solitary live oak and to the right (north toward Lipan Flat) may be seen some scattering of mesquite. (At South Concho) . . . crossed here easily after cleaning the brush from the banks. (At Dove Creek) . . . crossing effected without any difficulty after cutting out the brush from its banks. Both of these streams have heavy timber immediately on their banks, but no farther . . . grazing is only tolerable, the grass being old and dry. Pecan timber of large size is found."*

After 2 July 1849

SITE: Northwest toward the Middle Concho River, crossing Spring Creek and passing Lopez Peaks (Green Mounds of early travelers).

ACCOUNT: *"After passing Green Mounds the country becomes rolling prairie. The grass all along our route today and yesterday appeared dry and burnt off offering but little sustenance to our animals."*

Lt. F. T. Bryan records the grazing availability and condition in each of his accounts. In almost all cases the grazing is "excellent" or "good," and only tolerable when it appears recently "burnt off," possibly from wildfire. It is interesting to note that the date is 1849, and that bison are documented as abundant in the Concho Valley until the 1870s. Sheep were not brought into the valley until 1877, followed by longhorn cattle in 1880.

**John R. Bartlett**—Led the US-Mexican Boundary Survey team through the Concho Valley area on the way to El Paso. His route was similar to that of Lt. F. T. Bryan for most of the way. However, his description of the "flat prairie" between Kickapoo Creek and Spring Creek indicates that he was a few miles north of Bryan's route and on Lipan Flat.

22–23 October 1850

SITE: Westbound between Brady's Creek and Kickapoo Creek.

ACCOUNT: *"The country today has been flat . . . few trees except the mesquite—now and then a little mot* [sic] *of live oak was to be seen. The prairie dog colony continues."*

SITE: After passing Kickapoo Creek and traveling westward approximately six miles.

ACCOUNT: *"The hills were entirely barren of trees and shrubs, and as the stunted grass had recently been burnt off, the aspect before us and on both sides was one of extreme barrenness."*

SITE: After crossing the South Concho River and heading for Dove Creek.

ACCOUNT: *"Continue on our journey over a flat prairie interspersed with stunted mesquite."*

SITE: Camped on Spring Creek, then proceeding west on Spring Creek and striking "due west" to Lopez Creek.

ACCOUNT: *"The country traversed today has been flat and barren, no wood seen but stunted mesquite. Habitation of prairie dogs continued along our route."*

SITE: At Lopez Peaks (Irion County).

ACCOUNT: *"Ascended the eastern one, the sides stony and barren. Had recently been burnt over, hence their color was far from being green."*

SITE: Proceeding on to the Middle Concho River, describing the river bottom.

ACCOUNT: *"In several places where I could reach its shores through the thick brush and down a steep bank of twenty feet, I found the water in the middle of the stream less than two feet deep, its width was from twenty to thirty feet. As far as the eye could see on both sides the country was barren . . . no wood but mesquite. The river bottom scarcely extending beyond its immediate banks."*

SITE: Proceeding westward about twenty-five miles on the Middle Concho River bottom, passing Kiowa Creek.

ACCOUNT: *"The river or rather the creek followed today, ran through a valley quite barren save in its immediate vicinity. Beyond the valley there was a gradual slope to the hills, which extended on both sides of the creek for its entire length. Not a tree or shrub grew on these hills which seemed to be covered with stones and rocks."*

**Waterman L. Ormsby**—These are Ormsby's accounts as he traveled from Fort Chadbourne through Grape Creek, crossed the North Concho River, and eventually came to the head of the Middle Concho River.

23 September 1858

SITE: Arriving at Fort Chadbourne.

ACCOUNT: *"A few hours ride brought us to Chadbourne, a military station on a bend of the little Colorado River, exactly on the thirty-second parallel of latitude. . . . The most direct course to El Paso would be from this point along the thirty-second parallel, but the much dreaded Llano Estacado, or Stalked* [Staked] *Plain, interposes its waterless barrenness, and our course must still be in a southwesterly direction to the head of the Concho River, a tributary of the little Colorado, and thence to the 'Horsehead Crossing' of the Pecos River, taking us a degree further south, which we have to regain by following up the Pecos—all of which might have been saved had the money which has been expended in trying to sink artesian wells on the Staked Plains been applied to the purpose of building plain tanks to catch the water when it falls, as often it does in copious quantities."*

SITE: Leaving Fort Chadbourne.

ACCOUNT: *"Fortunately our course was a clear and straight one, lead-*

ing across an apparently boundless prairie, with not a tree or shrub to be seen, the parched grass almost glistening in the light of the moon. The night was clear and bright, the road pretty level, and the mules willing, and I soon ceased to regret having started."

The occasional mention of "not a tree or shrub to be seen" on "prairies" and "not a tree or shrub" growing on hill slopes, even though "covered with stones and rocks," indicates that at least some areas were nearly devoid of woody plants of any kind, or so recently burned that regrowth was not visible. From this, we must conclude that the Concho Valley was diverse in its appearance; some areas were completely open grassland, other upland areas had considerable woody growth, and most areas had varying mixtures of woody and herbaceous communities. However, compared to that of today, the landscape was predominantly open.

"But about 2 a.m. we came to a steep and stony hill, obstinately jutting from the prairie, right in our path and impossible of avoidance. One mule could neither be coaxed or driven up, so we had to camp until morning, when after much difficulty, we ascended the hill."

24 September 1858

SITE: Arriving at Grape Creek Station.

ACCOUNT: "The station was near Grape Creek, a fine stream, and also near some fine timber—two desirable things not to be found everywhere in Texas. The distance from this point to the head of the Concho River being fifty-six miles, and there being no inhabited station between."

25 September 1858

SITE: Arriving at an unnamed camp along the Middle Concho River, and then traveling west to the head of the Middle Concho River.

ACCOUNT: "We were off once more at a good pace. Our road lay over the rolling prairies studded with mesquite timber. A few miles from Grape Creek we crossed the [North] Concho, and then leaving the old road, which follows its winding course, we took a new road, across the country, which has been made under the supervision of the company—a ride of about thirty miles, the new road being very passable. We strike the [Middle] Concho again at a station about twenty-five miles from Grape Creek and fifty-five miles from Chadbourne, after following the [Middle] Concho to its source on the borders of the dreaded Staked Plain."

Descriptions of the Concho Valley up until this time indicated the dominance of open, sparsely wooded grassland on uplands, with wooded stream banks and bottoms. Shortly thereafter, woody encroachment increased, namely that of mesquite (*Prosopis glandulosa*) and juniper (*Juniperus* spp.). It is important to note that mesquite and juniper are native species. Also noteworthy is that mesquite is mentioned in these accounts starting in 1849, but juniper (commonly referred to as cedar) is not mentioned.

The Jumano Indians are cited as hunting bison and gathering pecans in the Concho Valley as they wintered along the Rio Grande in the late seventeenth century (Maxwell 1979). Fort Concho was established in 1867. Bison were documented as abundant in the Concho Valley until the 1870s. Sheep were first brought into the Concho Valley by John Arden in 1877 (Hunt and Leffler 2010). Longhorn cattle were brought into Irion County by Billy Childress in 1880 (Hunt and Leffler 2010). Cattle numbers in Texas increased from about 4,600,000 head in 1873 to 9,334,000 in 1902. In 2013, Texas beef cattle numbered around 4,015,000, a dramatic decrease to pre-1873 populations. I attribute these lower numbers at least partly to more landowners and managers becoming better stewards of the land.

Several authors as well as local culture attribute the increase in mesquite and juniper density in the Concho Valley and Texas to two main factors: the suppression of grassland fires and the introduction of livestock. There is evidence to suggest that fires were viewed negatively and ceased in the area shortly after the accounts given above. A surgeon stationed at Fort Concho wrote in 1869: "Malicious persons burned the prairie grass from around the post in every direction for more than fifty miles" (Maxwell 1979). Another possible cause of decreased fire frequency was the absence of good, continuous fine fuel, or grass, which would have resulted from the high cattle numbers. Some of the accounts given above describe the area as "recently burned off" prior to their arrival. Dr. Butch Taylor writes of other accounts from throughout Texas that have parallels to the accounts given here that mention arriving on previously burned prairie and grasslands.

I would also suggest that livestock alone were not a problem, but these livestock numbers, introduced by a new ranching culture, remained well over the carrying capacity. Regarding the Edwards Plateau, Dr. Taylor wrote, "For the first time in human history, a new culture of people were ranching in a semi-arid environment without any ancestral experience to help guide their management decisions."

## Overview of Livestock Grazing Management

As a rangeland specialist, I have worked with many landowners and managers and assisted them with grazing management plans over the years. These grazing operations were diverse in the stock involved—including cattle, sheep, goats, horses, exotics, or all of the above—as well as in their ownership, their management, their terrain, and their goals. I have also worked in some of the driest years as well as some of the wettest.

Good management principles are the same for every operation in this semiarid region. I will share some of those that I feel are the most important based on my experience.

- Stocking rate is the most important decision you can make on your ranch, year in and year out. Be flexible; your stocking rate may change often.
- Every pasture needs a rest and rotation system. In my experience and in the experience of many ranchers who have worked with me, rotation is always better than continuous grazing. All plant communities under grazing pressure need a rest at some point. This is the natural way. In semiarid west central Texas, more rest is usually required combined with shorter grazing periods.
- Consider *all* animals in calculating your carrying capacity. In developing a grazing plan, ranchers would often state only the number of cattle. Horses were often overlooked, or dismissed as "just staying in the horse pasture." The same goes for llamas, donkeys, and dwarf goats—they are all consumers of forage and need to be accounted for.
- All grazing animals will continue to graze if given the opportunity. Feeding hay does not change this.
- Grazing with native grasses in this region means you have a finite amount of forage per year. When a forage inventory is conducted and discounts taken, the result is what you have for the whole season. Think of it as the spring rains making a deposit in your "bank account" that you will have to make withdrawals from for the rest of the year. These native grasses cannot be expected to have quick regeneration as do the Bermuda grasses in more humid regions farther east.
- This is not Austin, or the Hill Country. Be critical of what you

read and hear. This is a semiarid region and the precipitation, plants, and associated management are different. An example is managing "cedar." Cedar is predominantly Ashe juniper (*Juniperus ashei*) in the Hill Country; you cut it down at the base and it does not grow back. The "cedar" in the San Angelo region is a mixture of Ashe juniper and redberry juniper (*J. pinchotii*); many areas are exclusively redberry juniper. Redberry juniper is a resprouter; cutting it down at the base only encourages resprouting and possibly tiller development.
- The ranchers who reduced livestock numbers or destocked early in drought years had more healthy and vigorous plant communities when the rains came back. These places were hands-down more resilient and bounced back more quickly than those ranches that did not destock or reduce numbers during drought.
- Know your plants so you know how to manage them.

Grazing animals eat approximately 3 percent of their body weight daily. Managing livestock on rangeland is a matter of balancing the supply (how much usable forage you have) with the demand (how many consumers you have and how much they are eating per day). Again, think of it as a savings account. You want to make only partial withdrawals from your savings account so you have enough left for the health of the plant and it will be there for you next year. Obtain the help of a rangeland or grazing specialist, who will help you determine a forage inventory and develop a grazing plan specific for your operation.

The following chart compares the forage intake of various types of animals. Since exotic ranches are increasing in the area, it is important to have consumption information on them as well. In determining a grazing plan it is necessary to have a unit of measure that can be used in calculations. This base unit of measure is the "animal unit." An animal unit (AU) is based on 1,000 lbs. of forage-consuming flesh. An animal unit equivalent (AUE) is a numerical figure expressing the forage requirements of a particular kind and class of animal relative to the requirements for one animal unit.

## Animal Unit Equivalent Chart for Domestic Livestock, Native Wildlife, and Exotic Wildlife Found in the Concho Valley and West Central Texas

| Kind of animal | Average body weight (lbs.) | Average daily consumption, (% of body weight, air-dry forage) | Annual forage consumption (lbs., air-dry forage) | Number of AU per head | Number of head per AU |
|---|---|---|---|---|---|
| **Domestic livestock** | | | | | |
| Beef cattle (cow)* | 1000 | 3.0 | 10950 | 1.00 | 1.0 |
| Horse | 1100 | 3.0 | 12045 | 1.10 | 0.9 |
| Sheep, meat and wool (nanny) | 130 | 3.5 | 1661 | 0.15 | 6.6 |
| Spanish goat (nanny) | 80 | 4.0 | 1168 | 0.11 | 9.4 |
| Boer goat (nanny) | 130 | 3.5 | 1661 | 0.15 | 6.6 |
| Boer x Spanish goat (nanny) | 100 | 3.5 | 1278 | 0.12 | 8.6 |
| Angora goat (nanny) | 70 | 4.0 | 1022 | 0.09 | 10.7 |
| **Native wildlife**** | | | | | |
| White-tailed deer | 100 | 3.5 | 1278 | 0.12 | 8.6 |
| Mule deer | 135 | 3.5 | 1725 | 0.16 | 6.3 |
| Pronghorn antelope | 90 | 4.0 | 1314 | 0.12 | 8.3 |
| **Exotic wildlife**** | | | | | |
| Axis deer | 150 | 3.5 | 1916 | 0.18 | 5.7 |
| Sika deer | 145 | 3.5 | 1852 | 0.17 | 5.9 |
| Fallow deer | 130 | 3.5 | 1661 | 0.15 | 6.6 |
| Elk | 800 | 3.0 | 8760 | 0.80 | 1.3 |
| Red deer | 350 | 3.5 | 4471 | 0.41 | 2.4 |
| Barasingha deer | 350 | 3.5 | 4471 | 0.41 | 2.4 |
| Sambar deer | 400 | 3.5 | 5110 | 0.47 | 2.1 |
| Père David's deer | 400 | 3.5 | 5110 | 0.47 | 2.1 |
| Mouflon/Barbados sheep | 120 | 3.5 | 1533 | 0.14 | 7.1 |
| Aoudad sheep | 150 | 3.5 | 1916 | 0.18 | 5.7 |
| Sable antelope | 500 | 3.0 | 5475 | 0.50 | 2.0 |
| Blackbuck antelope | 75 | 4.0 | 1095 | 0.10 | 10.0 |
| Nilgai antelope | 350 | 3.5 | 4471 | 0.41 | 2.4 |
| Scimitar-horned oryx | 400 | 3.5 | 5110 | 0.47 | 2.1 |
| Gemsbok oryx | 400 | 3.5 | 5110 | 0.47 | 2.1 |
| Arabian oryx | 150 | 3.5 | 1916 | 0.18 | 5.7 |
| Addax | 250 | 3.5 | 3194 | 0.29 | 3.4 |
| Ibex | 100 | 3.5 | 1278 | 0.12 | 8.6 |
| Impala | 130 | 3.5 | 1661 | 0.15 | 6.6 |
| Common eland | 1000 | 3.0 | 10950 | 1.00 | 1.0 |
| Greater kudu | 450 | 3.5 | 5749 | 0.53 | 1.9 |
| Sitatunga | 200 | 3.5 | 2555 | 0.23 | 4.3 |
| Waterbuck | 500 | 3.0 | 5475 | 0.50 | 2.0 |
| Thomson's gazelle | 85 | 4.0 | 1241 | 0.11 | 8.8 |
| Llama | 325 | 3.5 | 4152 | 0.38 | 2.6 |
| Zebra | 650 | 3.0 | 7118 | 0.65 | 1.5 |

*Note:* This chart is based on the standard concept of an animal unit being one 1,000 lb. beef cow consuming an average of 3 percent of her body weight (air-dry forage) daily throughout her yearly production cycle. Actual daily consumption will vary considerably throughout the year. Young of the year (calves, lambs, kids, fawns) are considered part of the maternal animal until weaning. After weaning, they are considered a separate animal and should be added to the herd.

*Other sizes and classes of cattle are usually calculated as 0.1 AU per 100 lbs. of body weight. (So a 700 lb. steer = 0.7 AU; a 1,200 lb. cow = 1.2 AU; and a 1,500 lb. bull = 1.5 AU.)

**For wildlife species, the average body weight and AU equivalent is based on a normal population consisting of males, females, and yearling animals. For populations that vary from this, appropriate adjustments can be made.

# Reading the Landscape in West Central Texas

Land stewards or managers should know how to "read" the land that they are managing. Reading the landscape is the ability to gather information from visual and sensory indicators in order to make informed decisions. The "language" that we use in our reading is our plants: the ability to recognize them, manage them, and know how they respond to certain management.

## Browse Lines

Browse lines develop on taller shrubs and trees when animals remove the branch, twig, and leaf growth within their reach. The photos show a live oak community and a hackberry tree. The exposed tree trunks and raised canopy height are *not* a natural growth form of our native tree species. The height of the browse line varies with the animal. White-tailed deer normally browse only about 3 to 4 feet off the ground. Goats, which can browse while standing on their hind legs, can browse lines up to 4 to 5 feet tall. Some large exotics can browse to a height of 6 feet. While some trees may remain healthy after a browse line is established, the available browse becomes out of reach and therefore of no more use to the animal. If periodic rest or deferment from livestock does not occur and correct deer popu-

Live Oak Community—This live oak motte has a severe browse line, along with the familiar sight of bare ground, rocks, and prickly pear. Ironically, the only "good grass" and forbs that may be present may be protected by the prickly pear. Unfortunately, this is a common occurrence with goat producers who do not practice good grazing management. Photo by George Clendenin, USDA-NRCS.

Hackberry— This tree has a severe browse line. There still appears to be grass on the landscape, which may indicate excessive goat or white-tailed deer numbers. Hackberry is very desirable tree browse for goats and white-tailed deer; therefore this lone tree in the open is an easy target. Photo by George Clendenin, USDA-NRCS.

lations are not maintained, then the ability of these plants to reproduce from seeds will also be eliminated. If this happens, these trees will continue to mature and there will be no new desirable trees to replace them when they die. In such cases, the better browse plants are often replaced by low-quality brush.

## Hedging

Hedging is the growth form of woody plants that develops over time when the tips of twigs and leaves are consumed, thereby causing more lateral twigs and leaves to develop. Moderate hedging is not harmful to a plant and can actually be beneficial since more leaves and twigs are stimulated to grow. Moderate hedging may make a plant look like a small umbrella-shaped shrub, with adequate leaf

Ephedra—This is a good species with which to evaluate the degree of use in a pasture. It is browsed year-round by goats, sheep, and deer but is of greatest value in winter, when it may be one of the few green plants available. The ephedra pictured here has experienced a light to moderate degree of hedging but still appears healthy and thriving, with many herbaceous "leaves" still available. Photos by George Clendenin, USDA-NRCS.

area accessible to the browsing animal. Severe hedging, however, creates a network of stubby twigs without enough leaf area or production. Severe hedging hurts plant health and vigor and may eventually lead to the death of the plant. Test your desirable shrubs for moderate or severe hedging: pat your open, flat hand over a shrub. Does it feel soft and springy, with adequate leaves? Or does it feel "pokey," with your hand hitting only a network of stubby twigs? This test combined with visual observation may help you determine the degree of hedging of a particular woody species.

## Mottes

Mottes, or groupings of woody species, are a valuable west central Texas rangeland resource and form a sort of West Texas oasis. Even

A quick survey of this motte revealed over 10 woody species! Included are skunkbush, algerita, littleleaf sumac, juniper, hackberry, flameleaf sumac, green condalia, Texas persimmon, bumelia, and tickle tongue. Photo by George Clendenin, USDA-NRCS.

a small motte can contain various woody species that offer habitat, forage, protection, and shade for various wildlife species. These mottes form a microclimate that offers cooler temperatures, greater soil moisture, and increased water infiltration than on adjacent open rangeland.

## Resprouting

Some woody plants naturally produce a network of new growth from the root system. This maintains the health and vigor of the plant but also provides accessible forage to browsing animals. Root resprouting of a plant can also be stimulated after a disturbance. Some typical disturbances include fire, animal browsing, and mechanical removal such as shredding or top cutting. Resprouting can be good or bad depending on the value of the plant. Most woody plants resprout in the Concho Valley, with the exception of Ashe juniper.

Redberry juniper and mesquite both resprout after top removal; therefore care must be taken not to stimulate regrowth if the goal is to kill or remove the plant. Other desirable browse plants such as shin oak species can be encouraged to stimulate new growth to provide accessible forage.

Shin oak species are prolific resprouters after a prescribed burn, offering palatable and accessible browse. The photo above was taken 5 months after a prescribed fire in south Tom Green County. Photo by George Clendenin, USDA-NRCS.

This photo shows redberry juniper resprouting after an extreme wildfire.
Photo by George Clendenin, USDA-NRCS.

### Plant Identification through Smell

Many times you can smell the aroma of a mysterious plant after traveling through a pasture. Many of these aromas are released after the leaves are crushed, as when you walk or drive through the pasture. Some are aromatic at the blooming stage and merely need the wind to carry the scent. The following list will give you clues to the identity of some of these aromatic plants. My thanks to Jake Landers for providing some of these descriptions.

**Minty Smell**
false pennyroyal (*Hedeoma drummondii*) — strong minty smell
common horehound (*Marrubium vulgare*)

**Sage Smell**
Mexican sagewort (*Artemisia ludoviciana*)
mealycup sage (*Salvia farinacea*) — minty sage
lance-leaf sage (*Salvia reflexa*) — minty sage

**Citrus Smell**
Dutchman's breeches (*Thamnosma texana*) — citrus peel
tickle tongue (*Zanthoxylum hirsutum*) — citrus peel

**Sweet and Sour Smell**
Huisache daisy (*Amblyolepis setigera*) in bloom
whitebrush (*Aloysia gratissima*) in bloom — sweet honey

**Unusual Odors**
Ashe juniper (*Juniperus ashei*) — camphor smell
redberry juniper (*Juniperus pinchotii*) — "3-in-One" oil smell
skunkbush (*Rhus trilobata*) — light skunky odor when leaves are crushed
buffalo gourd (*Cucurbita foetidissima*) — awful smell, stays with you if you come in contact with this plant
broom snakeweed (*Gutierrezia sarothrae*) — sticky resin and turpentine odor
scarlet musk flower (*Nyctaginia capitata*) — musky smell
lemon beebalm (*Monarda citriodora*) — liniment on a sweaty horse
western soapberry (*Sapindus saponaria* var. *drummondii*) — scraped bark smells skunky

crotons (*Croton* spp.)—the crushed leaves of all crotons have a distinctive strong smell; don't taste—it can be overpowering

ragweeds (*Ambrosia* spp.)—very distinctive but hard to describe

wild onion (*Allium drummondii*)—the leaf and tuber have a strong onion smell

## Grass Establishment

Perhaps when we "speed read" the landscape, we may tend to miss the obvious. We spend a lot of time discussing brush and browse in west central Texas. Perhaps a more focused, foundational question for the land manager should be "Do I have a good grass stand establishment?" Whether you manage livestock or wildlife, or just like the view from the porch, grass is the foundation for rangeland.

A good blend of native grasses and forbs is needed on our rangelands, along with our native shrubs and trees. Perennial warm-season bunchgrasses have been shown scientifically to reduce water runoff, increase water filtration back into the rancher's soil profile, reduce erosion and soil loss, and provide forage and habitat for livestock and wildlife. Ranchers who wish to start prescribed burning on their land will also need good grass establishment and continuity, in order to provide a good fuel source for a successful burn.

Along stream banks, tall bunchgrasses with extensive, deep root

Photo by George Clendenin, USDA-NRCS.

systems "lie over" during a flood event, with their roots binding the soil, and return to their natural upright positions after the floodwaters subside. Agricultural engineers and researchers have examined tall native bunchgrasses (e.g., switchgrass) for their ability to prevent soil loss on engineered grassed waterways and spillways of dams and ponds.

With this many benefits, one can simply state: Grass is good.

## Managing Riparian Areas in West Central Texas

STEVE NELLE AND GEORGE CLENDENIN

Riparian areas are the long, narrow ribbons of land that lie immediately adjacent to perennial and seasonal creeks and rivers; they can also occur around springs. These are the areas that overflow during heavy rainfall and runoff and are often referred to as bottomlands. Riparian areas often have deep, rich soils deposited from many previous floods. These floodplain areas often have favorable moisture conditions much of the year because of the shallow water table.

One of the key attributes of a healthy riparian area is the presence of dense riparian vegetation, which is specialized to withstand frequent flooding. The root systems of riparian plants are much denser and stronger than the roots of upland vegetation. Research indicates that some riparian plants have 10 to 20 miles of roots in a cubic foot of soil. In the Concho Valley, good riparian vegetation consists of a mixture of trees, shrubs, grasses, sedges, and forbs. Some of the more common riparian plants include buttonbush, black willow, little walnut, elm, pecan, switchgrass, eastern gamagrass, bushy bluestem, knotgrass, spike rush, and Emory sedge.

The primary functional benefits of riparian areas are to dissipate the energy of floodwaters and protect the banks from excess erosion. When the energy of floodwater is dissipated by dense vegetation, the velocity is reduced and the damage is diminished. When water is slowed, some of the sediment drops and becomes trapped in the vegetation. Over time, this capturing of sediment builds larger and larger floodplains.

During periods of out-of-bank flooding, riparian areas store large volumes of water in the banks and floodplain. This is sometimes referred to as the "riparian sponge." Then, after the floodwater recedes, water begins to seep out of the sponge back into the channel to help

Photo by Mark Meyer, USDA-NRCS.

sustain base flow, even during dry periods. When proper vegetation is present in riparian areas, these beneficial functions will occur. If the vegetation has been altered or diminished, these functions will be greatly reduced.

Because of the presence of extra water and deep soil, riparian areas are often 3 to 5 times more productive than adjacent upland areas; yet they usually make up less than 5 percent of the total land area. On livestock ranches, this can create a problem. Since riparian areas offer abundant forage, water, and shade, they are often subject to concentrated livestock grazing. Even when a pasture is properly stocked with livestock, its small riparian area will often be heavily grazed. In some cases, this has caused the degradation of riparian vegetation and a reduction in the benefits it provides.

In order to maintain or restore good, dense riparian vegetation, many ranchers have chosen to fence their creek areas separately to allow specialized management. These riparian pastures are grazed once or twice each year for brief periods to take advantage of the large volume of forage. Then the livestock are moved to upland pastures so that the riparian pasture can rest and the vegetation can regrow. This type of grazing management is becoming more widely accepted and provides benefits not only to the ranch, but also to everyone downstream. For riparian areas that have been more seriously degraded, some ranchers choose to withhold grazing for 5 to 10 years to jump-start the recovery of desirable vegetation. Then, after a period of healing, the riparian pasture is once again grazed on a periodic basis as described above.

In some areas of the Concho Valley, excessive populations of deer and exotic wildlife are damaging riparian areas. Free-ranging axis deer have become numerous along portions of the North Concho River and are damaging desirable vegetation. High populations of white-tailed deer can also harm some riparian plants by overbrowsing. Their numbers should be controlled and managed through proper hunting to reduce these problems. Feral hogs are not yet common across most of the Concho Valley, but their range is expanding and their numbers are growing. Landowners and hunters are advised to aggressively control feral hogs and not allow them to get started. In areas where hogs are already present, aggressive shooting and trapping are needed to keep numbers suppressed.

Landowners who have creek or river frontage are encouraged to keep a buffer of dense riparian vegetation. Mowing, burning, plowing, driving, excessive use of chemicals, and excessive human or livestock use are discouraged in riparian areas. The results of good riparian management include improvements in water quality, fish and wildlife habitat, forage production, land value, and recreational and aesthetic value.

### Dove Creek

Healthy, functional riparian areas in the Concho Valley normally include a mixture of grasses, sedges, forbs, and woody plants. Specialized management is needed to maintain good riparian conditions. Photo by George Clendenin, USDA-NRCS.

A variety of unique and specialized plants are often found in riparian areas. Cardinal flower is especially beneficial for hummingbirds. Photo by George Clendenin, USDA-NRCS.

Creeks and rivers in the Concho Valley often have a combination of deeper pools and shallow riffles, providing good aquatic and fish habitat. Photo by George Clendenin, USDA-NRCS.

# South Concho River

The roots of trees and shrubs help protect banks and provide shade. Densely wooded riparian areas are extremely important for many species of wildlife. Common trees along the South Concho River include pecan, live oak, bur oak, red mulberry, American elm, and western soapberry. Photo by Jaime Tankersley, USDA-NRCS.

Conscientious, conservation-minded landowners provide the stewardship required to maintain healthy, intact riparian areas on the headwaters of the South Concho River. Photo by George Clendenin, USDA-NRCS.

Where It Begins—A spring bubbling forth clear water at the headwaters of the South Concho River. Photo by George Clendenin, USDA-NRCS.

## Spring Creek

A well-vegetated floodplain adjacent to creeks and rivers helps dissipate energy and trap sediment during flooding. Switchgrass is one of the most important riparian stabilizers in the Concho Valley. Photo by George Clendenin, USDA-NRCS.

Purple gayfeather is commonly found growing along creek banks in healthy riparian areas.
Photo by George Clendenin, USDA-NRCS.

Dense riparian vegetation helps armor and protect creek banks with strong, reinforcing root systems that resist erosion. Here, a combination of switchgrass, eastern gamagrass, and sawgrass help maintain a stable bank.
Photo by George Clendenin, USDA-NRCS.

# Middle Concho River

A good visual comparison of a well-vegetated riparian area on the left bank and sparse vegetation on the right. Improper grazing and/or excessive recreational activity can degrade riparian vegetation. Photo by George Clendenin, USDA-NRCS.

Large logs and fallen trees are important components of a healthy riparian area. This large wood helps dissipate energy and slow the movement of sediment during flood events. As the logs become partially buried, they remain intact for many years, helping reinforce channels and banks. Photo by George Clendenin, USDA-NRCS.

## Riparian Plants of the Concho Valley and West Central Texas

**Grasses, Sedges, Rushes**
Switchgrass (*Panicum virgatum*)
Eastern gamagrass (*Tripsacum dactyloides*)
Knotgrass (*Panicum distichum*)
Bushy bluestem (*Andropogon glomeratus*)
Rice cutgrass (*Leersia oryzoides*)
Hairyseed paspalum (*Paspalum pubiflorum*)
Rabbitsfoot grass (*Polypogon monspeliensis*)
Bulb panicum (*Zuloagaea bulbosa*)
Wildrye (*Elymus* spp.)
Western wheatgrass (*Pascopyrum smithii*)
Vaseygrass (*Paspalum urvillei*)
Bermuda grass (*Cynodon dactylon*)
Emory sedge (*Carex emoryi*)
Spikerush (*Eleocharis* spp.)
Bulrush (*Schoenoplectus* spp.)
Flatsedge (*Cyperus* spp.)
Torrey rush (*Juncus torreyi*)
Tapered rush (*Juncus* spp.)
Cattail (*Typha* spp.)

**Woody Plants**
Buttonbush (*Cephalanthus occidentalis*)
Indigobush amorpha (*Amorpha fruticosa*)
Black willow (*Salix nigra*)
Spiny aster (*Leucosyris spinosa*)
Little walnut (*Juglans microcarpa*)
Pecan (*Carya illinoinensis*)
Live oak (*Quercus virginiana*)
Red mulberry (*Morus rubra*)
American elm (*Ulmus americana*)
Chinquapin oak (*Quercus muhlenbergii*)
Bur oak (*Quercus macrocarpa*)
Hackberry (*Celtis* spp.)
Western soapberry (*Sapindus* spp.)
Willow baccharis (*Baccharis* spp.)
Sycamore (*Platanus* spp.)
Wafer ash (*Fraxinus* spp.)
Grape (*Vitis* spp.)

**Forbs**
Water willow (*Justicia* spp.)
Tall goldenrod (*Solidago canadensis*)
Scouring rush (*Equisetum* spp.)
Water primrose (*Ludwigia peploides*)
Watercress (*Nasturtium* spp.)
Mint (*Mentha* spp.)
Water hyssop (*Bacopa* spp.)
Water pennywort (*Hydrocotyle* spp.)
Smooth bidens (*Bidens* spp.)
Large buttercup (*Ranunculus macranthus*)
Cardinal flower (*Lobelia cardinalis*)
Tall aster (*Aster* spp.)
Bog nettle (*Boehmeria* spp.)
Frogfruit (*Phyla nodiflora*)
Late boneset (*Eupatorium serotinum*)

# The History and Use of Fire in Texas and the Edwards Plateau

DR. CHARLES "BUTCH" TAYLOR

> *Somewhere in the past one of our ancestors picked up a flaming brand, held it and eventually set the grass on fire for his purpose. He had long recognized the fact that animals congregated on such burns, for he was a hunter and gatherer. No other creature will hold a flaming brand but man.*
> — Ed Komarek

## Introduction

Texas rangelands evolved under periodic fire, grazing, browsing, and a highly variable climate. The years are marked by alternating frequent wet and dry periods. Therefore, vegetation that evolved under this disturbance regime is well adapted to change. In fact, change (i.e., disturbance) is the norm and without it, rangeland vegetation would not have been as fit and diverse as it was when our forefathers first moved into Texas to establish their early livestock production enterprises.

As the early livestock producers moved into the semiarid regions of the state, they found grassland savanna (i.e., land dominated by grass with scattered trees and shrubs). There has always been a constant battle between woody plants and grasses to dominate rangeland. In the past, grass was lightly grazed by native wildlife and, with fire as a constant friend, dominated a major portion of the landscape. However, with the development of the range livestock industry and the constant overstocking of the range, fires were suppressed. As a result, across most Texas rangelands woody vegetation has increased in abundance relative to grasslands.

Fires burned frequently on dense grasslands and shrublands where fuels accumulated rapidly. Steep, rocky, less densely vegetated sites burned less, serving as firebreaks until the right mix of weather and fuel loads provided optimum conditions for fire. Variation in plant communities, combined with variable weather and topography, created landscapes where fire burned in patches, providing a variety of fuels, fire intensities, and habitats for livestock and wildlife.

Prior to human intervention, grasslands occurred over a major expanse of the earth's land surface (approximately 45 percent; Copeland 1978). Natural fire (from lightning, volcanoes, etc.) was a frequent feature of these natural grasslands and savannas, which evolved under a system of grazing and browsing (herbivory), drought, and periodic fire. The characteristics of grasses that permitted them to survive extreme droughts also provided protection from fires. Important to grass survival are the organs beneath the soil surface (roots, rhizomes, etc.), which allow only dead aboveground biomass to be exposed during droughts, or at other times of the year when grasses are dormant (Gleason 1923).

## Prehistory—The Period Prior to Native American and European Settlement

The relationship between humans and fire has been long and intimate. Humans set fires in grasslands all over the world. This practice gave them some control over the game animals they hunted and also changed vegetation structure and composition. Scientific studies throughout the world have shown that fire maintains grasslands by suppressing woody vegetation and stimulating grasses and forbs (Anderson 1990). Our ancestors lived on the plains and savannas of Africa from approximately 3 to 4 million years ago. They foraged on the lightning-caused and other natural burns much like animals in Africa do today. It is obvious that fire is not a tool newly discovered for modern-day range management but is in fact one of ancient lineage. In order to understand the present and future use of fire, we need to understand its history.

Before there were roads, towns and cities, rural fire departments, livestock, and humans, natural summer fires must have been something to behold. Just imagine the fuel loads that built up in an above-average precipitation year and the consequences of a lightning strike starting a fire in July or August. The fire would start small but quickly spread because of the wind from the thunderstorm. Soon the fire would be large enough to create its own wind, sucking in oxygen to feed its appetite for more fuel. Flame lengths would reach into the trees from the lead fire. Firebrands would travel hundreds of feet into the air and start new spot fires ahead and to the sides of the fire front. Soon the horizon would be covered with smoke and particulate matter, with both being lifted high into the atmosphere, possibly enough to create a rainstorm, but one without enough moisture to extinguish the fire. The momentum of the fire would carry it across rivers and streams and over hilltops and through ravines. Hundreds of thousands of acres would be burned. At night the fire would slow down

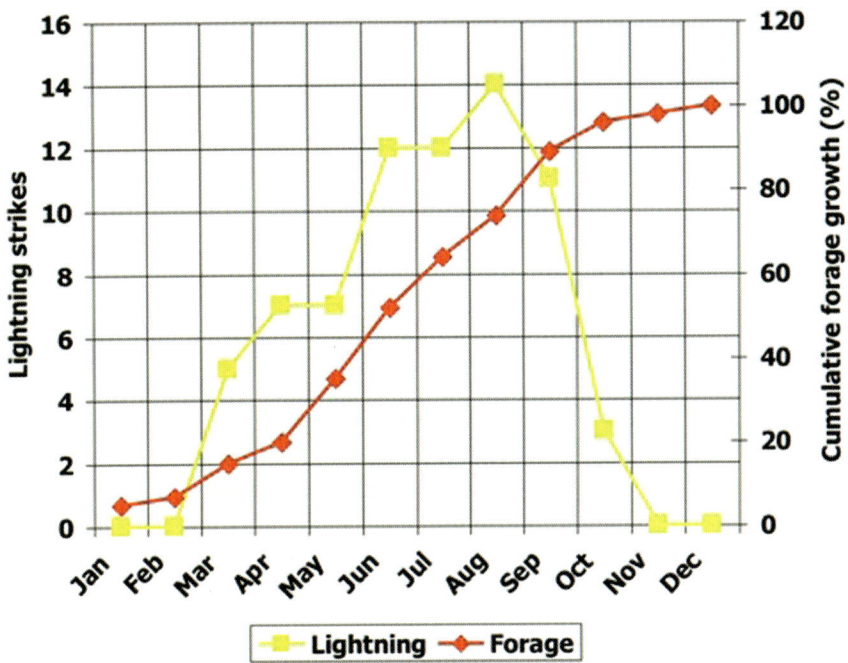

Figure 1. Relationship of lightning strikes to growth curve of forage.

and almost stop, as if it were resting. But the next day temperatures would rise, humidity would decrease, and winds would increase, and the fire would wake from its sleep and continue to spread across the landscape, seeking more fuel for its ravenous appetite. Depending on the weather conditions, the fire might burn for days or weeks; only nature would decide its fate.

In the fire's wake, untold acres of vegetation and litter would be burned down to mineral soil. The burned areas would look like a moonscape, charred and blackened, with no green leaf left for either ant or buffalo. With no soil moisture or rain, the landscape could appear uninhabitable for either human or beast for many months. But the rains would come and when they did, the perennial grasses with their energy and growing points stored underground would quickly reappear. The live oak, shin oak, and most other woody plants would also resprout from underground crowns or roots. In fact, if vegetation were sampled both before and after the burn, when plants had resumed their growth, species composition would almost be identical. For the western part of the Edwards Plateau, the few exceptions would be Ashe juniper, prickly pear, and tasajillo, which would be absent from the vegetative complex for some time.

It was extreme summer fires that historically maintained a major portion of the Texas landscape as grassland or savanna (fig. 1). The bulk of the warm-season grasses (i.e., over 70 percent) are generally grown by July or August. These two months are generally dry and hot, and they also correspond with peak lightning strikes and a high fuel load, a perfect system for frequent summer fire.

### The First Americans and Fire

When the first humans appeared on the North American continent via the land bridge from Asia, the plains were stocked with large species of animals such as the giant buffalo, mammoth, elephant, camel, and horse, which are now extinct (Martin 1975). Armed with only spears, darts, and later, bows and arrows, the hunters probably had difficulty bringing down these huge, tough animals. Their most effective hunting method was fire, which they used to stampede the animals over a cliff or into a ravine as well as to attract them to burned areas once the immature vegetation began to grow.

Eventually, for some unknown reason, all the species of larger animals disappeared. This left the country open to occupation by smaller animals. One of these was the modern buffalo, which arrived on the North American continent approximately 12,000 years ago. These buffalo multiplied and spread until at their peak they numbered about 40 million head and occupied nearly half of North America. These large numbers of buffalo had a profound effect on the life and culture of Native Americans.

Native Americans observed that buffalo consumed mostly grass and preferred areas dominated by grass. They also realized that fire was needed to maintain open prairies, especially in the eastern part of America. Fire was an important tool for manipulating the vegetation to improve buffalo habitat. Without fire, the alternative food source in potential forested areas was deer and squirrel meat. "In the spring the Indians throughout the buffalo country burned off the old grass in places where they had not used a fire drive in the previous autumn. Until the new grass attracted the buffalo herds, the Indians hauled in the animals drowned in the river during the winter" (Haines 1970).

### Important Events That Influenced Edwards Plateau Vegetation
- 5,000 years ago early humans were building burned rock midden sites in the Edwards Plateau for the processing of acorns and other food items.
- Humans continued to use fire and flint-tipped spears to hunt

game. They burned off the coarse grass to grow young tender forage, which attracted game. They also used fire to drive buffalo off cliffs and into canyons.
- About 1,400 years ago, bows and arrows were added as hunting tools.
- In 1519 Hernando Cortés brought 16 horses from Spain to Mexico. This may have been the first reintroduction of an extinct species back to the North American continent. In 1541 Hernando de Soto crossed the Trinity River with horses (Spanish Barbs) and within 250 years the Native Americans had developed their own horse culture.
- Horses greatly facilitated the hunting skills of the Native Americans and made their life much easier. Also, because of the horse, the Comanches prevented the Spanish from colonizing any farther north into the Edwards Plateau region than San Antonio.
- In 1821 Stephen F. Austin was given permission to settle his colony around the Brazos and Colorado Rivers in Texas. The Spanish granted him permission because they thought his colony would act as a barrier against the Comanches and provide some protection for the people living along the Rio Grande.
- In the 1840s German immigrants signed a treaty with the Comanches to settle around the Fredericksburg area.
- In the 1860s President Lincoln signed the Homestead Act and granted more than 90 million acres of western rangeland to the railroads. This was the start of the elimination of the buffalo.
- In the 1870s the buffalo were essentially eliminated from the range. This ended the dominance of Indians in Texas and facilitated the development of the livestock industry. Early pioneer ranchers saw a rangeland that had been devoid of buffalo grazing for a number of years, which led to an overestimate of the carrying capacity.
- In 1884 Texas passed a law making the burning of grass a felony.
- Because of many negative experiences with fire, much of the ranching industry implemented fire suppression techniques rather than prescribed fire. For example, most large ranches began to use fireguards. The XIT ranch began plowing

fireguards in 1885, the first year cattle were placed on its range. Within a year, over 1,000 miles of fireguards, each 100 feet wide, had been plowed up on the ranch.

### Early Anglo Texans and Fire

Two major changes occurred across Texas and much of the western United States from about 1700 to 1900. Grazing by free-roaming animals such as buffalo changed to grazing by relatively free-roaming livestock, and ultimately to confined livestock (Webb 1931). Concomitant with this change was the influence of early settlers on the frequency, timing, placement, and extent of fires.

> One certain indicator of the importance of prairies to early settlers in eastern Texas was their practice, observed in even the earliest years of colonization, of regularly setting fire to the grasslands. This practice, perhaps inherited from the Indians, supposedly destroyed weeds and dead grass, so as to make room for the new grass. In addition, the fire removed all bushes and young trees, thereby preserving the prairie from encroachment by the forest. Indeed, it is fairly well agreed that most of the prairies in the eastern half of the United States were created by Indian clearing activity and maintained by annual firing. For the Indian, preservation of prairies meant that grazing bison would remain in the area, while for the Anglo-American, annual burning of the grasslands assured ample forage for the herds of cattle typically owned by the frontiersmen. (Smith 1899)

Amos Parker reported in 1834 that "prairies are all burnt over twice a year—in mid-summer, and about the first of winter" (Parker 1836).

Reports of early American pioneers exist, but it is important to remember that the historical record is not consistent and is sometimes contradictory about the vegetation that existed at the time of European contact (Smeins 1980). The historical records of central Texas make it apparent that the vegetation was highly variable, ranging from closed-canopy woodlands to open grasslands (Smeins 1980).

The sum of these accounts suggests that woody plant abundance was greatest in the eastern and southern parts of central Texas, where rivers and creeks drain and form steep canyons separated by high divides (Balcones Canyonlands region). To the north and west on the divide portion of the Edwards Plateau, there was much less woody vegetation, and more of a savanna or grassland dominated the landscape. Below are sample descriptions of the central Texas landscape from early pioneers.

Walter Prescott Webb wrote in his book about the Texas Rangers, "The whole country from the Llano via the head of the Guadalupe and Frio to the Nueces had been burned, and there were few places where water and grass could be found together. 'I have traveled a whole day at a time without finding any grass' according to Major John B. Jones in early Sept 1874" (Webb 1935).

In the early 1880s fire was an annual summer event in the western region of the Edwards Plateau. "As often as not the fire ran uncontrolled, destroying a million acres of grass down to the roots and wiping out the herder's plans to graze his flock through the winter" (Carlson 1982).

In 1854 Frederick Olmsted commented on the suppressed growth of live oaks near Austin: "The live oaks are often short, and even stunted in growth, lacking the rich vigor and full foliage of those further east. Occasionally a tree is met with which has escaped its share of injury from prairie burnings and northers, and has grown into a symmetrical and glorious beauty. But such are comparatively rare" (Olmstead 1857).

Based on his own observations, Julius Froebel discussed the relationship between fire and the abnormal age grouping of the honey mesquite population in West Texas in 1853.

> One peculiarity is the repeated occurrence of dead mesquite trees, of considerable size, with the growth of young ones,— there being no intermediate stage of size or age. This probably has been caused by repeated prairie fires, which destroyed the old trees, and prevented the growth of fresh ones....At Chihuahua a man who had been a great deal into this locality told me that for a long period no Indians had lived there, during which it was covered with a thick mesquite wood. Subsequently, certain hordes came here, and with them the prairie fires began. In later times the advance of the whites into Texas has driven back the savages, and restrained their visits; and the prairie fires ceasing, trees and shrubs have again appeared. It is asserted that this process may be watched throughout West Texas. (Froebel 1859)

As Andrew Gray wrote in 1856 about the lack of woody plants in central Texas, "Much of the soil is good, and I question if the grass set on fire annually by the Indians...together with the 'Northers,' which sweep with such violence over the plains are not to a great degree causes for the total absence of timber" (Gray 1856).

In 1923, V. L. Cory became a botanist at the Texas Agricultural Experiment Station at Sonora, where he met an elderly cowboy who told of driving a flock of sheep from Junction, Texas, through the area of the Texas A&M AgriLife Research Station and on to Juno 50 years before (circa 1873), when the vegetation was much different (Cory 1949). Cory reported the cowboy's description as follows: "There were no fences, nor was the country timbered then, as is now the case. This valley and all the other valleys then were free from woody plants; and the entire country was a prairie of tall bunch-grass, reaching at least to ones stirrups. . . . The only short-grass was around water holes and in depressions in the valleys." When asked about the abundance of juniper 50 years before, he remembered the cowboy replying: "These were few in number and confined to the headers [the gully or ravine-like beginnings of the branches of draws, or drainage courses, on the escarpment bordering the valleys] . . . but now the tall grass has gone, trees have spread everywhere, and the valleys, once having grass only, now are occupied chiefly by weeds, thorny shrubs, and prickly pear."

With the development of the range livestock industry and the constant overstocking of the range, fires were suppressed. Across most Texas rangelands woody vegetation has increased in abundance relative to grasslands (Smeins 1984; Archer 1994). This scenario is also accurately described by Foster (1917):

> The causes which have resulted in the spread of timbered areas are traceable directly to the interference of man. Before the white man established his ranch home in these hills the Indians burned over the country repeatedly and thus prevented any extension of forest areas. With the settlement of the country grazing became the only important industry. Large ranches in time were divided into smaller ranches and farms with a consequent fencing of ranges and pastures. Overgrazing has greatly reduced the density of grass vegetation. The practice of burning has during recent years, disappeared. The few fires which start are usually caused by carelessness, and with alternating wooded and open spaces and the close cropped grass, they burn only small areas. These conditions have operated to bring about a rapid extension of woody growth. Almost unquestionably the spread of

timbered areas received its impetus with the gradual disappearance of grassland fires.

Before the days of barbed wire fences, ranchers moved toward the western part of the state in search of grass for their livestock. They faced many obstacles such as hostile Indians, predators, adverse weather, and so forth. Initially the grass was free; however, eventually this caused problems, too, with the advent of barbed wire. In the early 1880s, grass was burned in retaliation for alleged grievances against the ranchers who were fencing the range. In 1884, for the second time, Texas passed a law making the burning of grass a felony (Haley 1929). It would be interesting to see whether this law is still on the books! However, it is ironic that these laws were passed with the idea of protecting grass; actually they had just the opposite effect.

Despite the legislation, large fires continued to sweep across western Texas during the early development of the livestock industry. The potential for fire was greatest during periods of dormancy or drought. An interesting story was reported by the *Crosbyton Review* on February 29, 1912:

> A very destructive fire occurred during the month of June, 1879. The fire originated on the Z-L Ranch in Crosby County, where there was considerable shinery. Hundreds of wild hogs ranged this dwarf oak country, prolific and hardy upon the acorns that grew there. Hank Smith, the first settler in the South Plains region, described this fire and the hogs. One day a cowboy decided he would set fire to the shineries and run the hogs out. He did it all right, but it is to be hoped that no one else will ever try to drive wild hogs out of a shinery country with fire. The fire got away and started on a wild rampage in a northeasterly direction. No one has ever learned for certain which way the hogs went. The fire swept the country now occupied by Crosbyton, Emma, Ralls, Lorenzo, and spreading as it went sped across the Blanco [Canyon] moving before a terrific wind from the southwest. At that time there was practically no cattle in the country, and few people cared where the fire went or what it did. Crossing the Blanco on it went into the Quitaque, Boggy Creek, North and South Pease River and Tule Canyon country, while before it fled and swarmed countless thousands of antelope, turkeys, hundreds of deer and a sprinkling of cattle and horses. The fire swept thousands of square miles of country to the south and south-

west, north and northeast of Mount Blanco. All through the country at that time, especially along the streams, were hundreds of magnificent groves of fine timber, particularly cottonwood and hackberry. This fire killed the timber and in effect literally wiped it out.

### West Texas Ranchers

By the early twentieth century, immigrants from Europe and pioneers from the United States had carved vast ranches out of the West Texas landscape. Most of the early ranchers were of European ancestry and moved onto the semiarid grasslands from areas with higher rainfall. For the first time in human history, a new culture of people was ranching in a semiarid environment without any ancestral experience to help guide their management decisions. It took them many years, some good and some bad, to learn how to ranch in Texas. Thus, these ranchers were practicing the "art" of grazing management, meaning that they observed management conditions and were able to learn from them. This knowledge has been passed from generation to generation in those families that still ranch today.

Since most of the western Edwards Plateau lacks surface water, the livestock industry was developed later here than in other parts of Texas. The lack of surface water also had an effect on the kind of ranching. For example, the early occupation of this part of the plateau by ranchers was characterized by a lack of organized government or written law. In the absence of such government, the early ranchers developed a body of unwritten rules of human action known as "the law of the range." This meant that no rent should be paid for grazing rights. It did not mean free grass in terms of common property, which anyone could use at will. It meant that the grass was free to the first individual who secured the range by getting there first. This arrangement was quite different from what was practiced in New England, where every citizen had the right to graze livestock on the "commons."

For most areas of Texas, livestock grazing was the pioneer industry. Cattle were usually brought in first, followed by sheep; however, this order was reversed in the western areas where there was no surface water and the range was some distance from rivers and streams (sheep require much less water than cattle). Because this section of the plateau had no running streams, ranges along the Conchos to the north, the San Saba to the northeast, and the Llanos to the east and southeast were stocked first. When sheepmen began to crowd each

Figure 2. Excessive stocking rates reduced the fuel load and fireproofed the rangeland. Picture taken early 1940s (Texas A&M AgriLife Research Station–Sonora).

other along these streams, especially during dry periods, the drier sections of the plateau were used for emergency feed.

One important change in the use of these sheep-grazing lands occurred in the late 1880s with the development of windmills. The section of land on which a well was located was either leased or bought from the state. Under this arrangement, the sheep were divided into bands of about 1,500 each and placed under a herder. Sometimes one well was held in partnership among 3 or 4 sheepmen. This type of management led to severe deterioration of rangeland around the wells.

Around 1910, ranchers started building wolf-proof fences. These were constructed of woven wire with 6-inch mesh, 42 to 52 inches high, attached to cedar posts, with a barbed wire on the ground (sometimes one on both sides of the posts at ground level), and 2 or 3 barbed wires above the woven wire. Wolf-proof fences, windmills, and water storage tanks increased the efficiency of ranching significantly; however, these technologies also allowed the early ranchers to continuously stock pastures at heavy rates. Eventually, this heavy grazing pressure reduced the midgrasses and fireproofed the rangeland (figs. 2 and 3).

## Fire Culture

One obvious trait of most of these early ranchers was the lack of a fire culture. A culture is something (such as a belief, practice, custom, etc.) that is passed from one generation to the next. With a few exceptions within the state, there has been very little evidence that early Texas ranchers used fire on a routine basis. However, I discovered one of the exceptions to the lack of fire culture on a trip to Leakey, Texas, in the winter of 2011. The purpose of the trip was to discuss

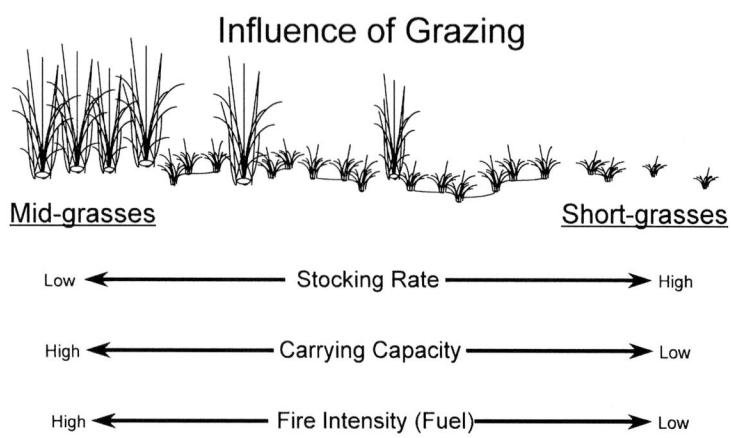

Figure 3. Short-grass production is not sufficient to provide adequate fuel loads for effective burns. Lack of fire allows cedar to encroach on Edwards Plateau rangelands.

the benefits of prescribed burn associations and, if the people at the meeting were interested, to help establish a prescribed burn association to serve the area. Approximately 30 ranchers were present at the meeting, and during the presentation and discussion of the benefits of prescribed burn associations, two individuals seemed to have strong ideas about prescribed burning. After the meeting I made a point to visit with both and received a magnum load of information on the proper use of prescribed fire. These two older gentlemen were third-generation ranchers who had learned to burn from their fathers and grandfathers. I asked them how their ancestors had learned how to burn and they quickly answered, "From the Indians." Below are some of their comments on prescribed burning that I find extremely interesting:

- "Controlled burning is a lot of work, but you have four times the production of grass at a cost you can afford."
- "Our land is less productive with our lifestyle of roads, permanent buildings, and too much TV watching."
- "Landowners cannot afford the cost of large machines . . . the end cost is much more if your tax money goes to a bureaucrat to do it for you."
- "Fire lanes [black lines] must be made in December at night, with a five mile per hour wind from the east. You are anticipating heavy dew by ten o'clock to put out the grass fire."
- "Seventy-five years ago I was given a burning split dry cedar root to poke at anything that would burn. It burned just like those drip pots [drip torches] you see on TV."
- "If you do not have a good black line, two people can move

into the wind. One person lights the fire and the other within five feet, puts out the side of the fire that is not wanted with a backpack water sprayer or most often a metal yard rake."
- "The controlled fire is made only in February from ten a.m. to two p.m. with a west wind under five miles an hour and humidity under twenty percent. This is always a back fire."
- "If all is not right, do not burn and waste your grass fuel. Generally only have zero to five days each year that are right to burn. If conditions are right today, you cannot wait for everyone. Burn!"
- "Most of our fences have roads beside them. Driving these roads many, many times while it is raining leaves a middle that can be burned. This and gravel bars have been our easiest and best fire lanes."
- "Burned roads are used for back fires in front of unplanned and unwanted fires."
- "Because of burn bans it is difficult to burn all of the roads today and this is why we have a lot more wildfires."
- "In the 1950s and 60s ranchers with four stock sprayers did what eight or ten fire trucks do today with much less water."
- "The animals will come to your burned area. They know there are not ticks or stomach worms and there is twenty percent more protein for two years."

Following the meeting, I also heard comments about estimating whether juniper would burn or not based on its color. I need to ask more questions about this as well as about other things that were discussed. To me it is obvious that this approach was developed before modern spray rigs, radios, and heavy equipment. Also, because this area is known for cedar posts, I wonder whether these fire prescriptions were not also developed to be used on areas that had recently been harvested and cleared by hand cutting cedar. Using back fires would have limited the acreage burned per day, but the ranchers may have conducted numerous days of burning as long as weather conditions held. It was quite an interesting meeting and ended in a burn association being started. Another interesting observation was that most of the old-timers (who burned using the Indian method) did not join the burn association. They were not against the idea of a burn association but did not see the need for one. They had also never heard of the Texas Commission on Environmental Quality (TCEQ).

Not too far from Leakey, another example of early use of fire was reported in Blanco County in the 1930s. This method was also directed at controlling juniper and involved a combination of hand cutting and fire. About 63,000 acres were cut and burned during 1938 alone in that single county (Jenkins 1939). The spectacular grass response on 65 acres that had been hand cut in the early summer of 1930 and then accidentally burned on an August afternoon was the "great stimulus in Blanco County that started the ranchers to cutting and burning their cedar," according to an article in *The Cattleman* in 1939 (Jenkins 1939). The reported benefits included an increase in livestock carrying capacities from 1 animal unit/20–30 acres to 1 animal unit/5–6 acres; springs that started flowing at much higher volumes, and some springs that began flowing that had not flowed in 45 years; a decrease in sheep losses from "blow fly" damage of 300 to 400 percent; reduced stress on livestock, horses, and mules from "blood sucker flies" (horseflies and deerflies); and an abundance of bobwhite quail in areas where they had not been present for 40–50 years.

### Current Use of Fire

Prescribed fire on the Edwards Plateau of Texas faces an uncertain future. Historical use of prescribed fire by ranchers was never widespread in this region; however, with the rapid increase in population and the increased urbanization of rangeland, air quality concerns, county burn bans, and so forth, the implementation of fire will be even more difficult in the future.

### Conclusions

Even though prescribed fire is a well-established management tool, current efforts to apply it to the Texas landscape are limited by economic, social, and legislative constraints. This presents a great challenge. Scientists and agency personnel can conduct research on the efficacy of management on rangeland ecosystems and identify the social perceptions that drive the application of that management, but these components simply make the need to incorporate those management strategies relevant. The issue at the heart of resource management is to identify and develop specific strategies that will empower and equip managers to overcome ecological, social, and economic constraints.

# Prescribed Burning in West Central Texas

Prescribed burning in the semiarid region of the Concho Valley is unique in its scale, application, and effects. It is important to understand the climate and the local plant community. High-intensity burns may take longer to recover in drier areas with less precipitation, such as the Concho Valley. The scale of burns here is much larger than in other, more populated parts of the state. It is not uncommon to burn one to several thousand acres in a single day, on a single burn plan.

There can be many objectives or goals in planning and conducting a prescribed burn. It is important to know which objectives are feasible for a region and which are not. Many landowners that I have worked with have attended a burn seminar or training elsewhere in the state and expect to apply the same information in the Concho Valley. These assumptions can be costly both financially and ecologically.

One example of this is that our "cedar" is not the same "cedar" as in the Hill Country of central Texas. Ours is mostly redberry juniper (*Juniperus pinchotii*), which is a resprouter. In other words, top-killing the plant with fire will not kill the plant. The "cedar" of the Hill Country is Ashe juniper (*J. ashei*), which will die if all of the green leaf matter is removed from the plant (by fire or mechanical means). Mesquite is also a resprouter. In fact, most brush species in the Concho Valley are resprouters. In the case of mesquite, a hot fire can make it worse, as mesquite will resprout more vigorously and develop more tillers, possibly turning it into more of a nuisance shrub than a single-stemmed tree.

## Top Four Objectives of Applying Prescribed Burns in West Central Texas

Different parts of the state will have different objectives based on their specific climate, plant communities, land use, public policy, and types of agricultural operations. Based on my experience with prescribed burning in west central Texas, I have summarized the most effective objectives specific to this region.

**Suppress prickly pear**—Prickly pear can be effectively suppressed and even killed by an effective burn. For prickly pear suppression, the hotter the fire, the greater the results. However, even properly applied cool-season burns can have long-term residual effects on prickly

Summer Burn— The author at work on a prescribed burn in Christoval, Texas. Photo by Dee Ann Littlefield, USDA-NRCS.

pear plant mortality. Tasajillo (*Opuntia leptocaulis*) is also effectively suppressed or killed by fire. Repeated application of fire over several years may be needed to gradually reduce cacti abundance.

**Suppress juvenile brush species**—Do not expect to kill mature juniper and mesquite. However, I advise landowners that juvenile plants may be killed or suppressed if the bud zone is at or near the surface. Additionally, even a cool fire through a mesquite savanna can help encourage grass growth under the canopy, which also provides competition to brush establishment.

**Increase desirable perennial grass communities**—Properly applied prescribed burning in west central Texas yields an increase in desirable grass communities. With properly timed moisture, there can be more grass several months after a burn than existed before the burn. Also, the plant communities that return are often healthier and more vigorous in production. Many plant photos in this book were taken after a prescribed burn.

**Manage fuel loads**—This is often overlooked by ranchers and the general public. Ranchers who have installed infrastructure such as fireguards and who have applied burns are typically better equipped and have better-managed fuel loads in the event of wildfire. A prescribed burn rotation on rangelands mitigates "ladder fuels," or brushy species under taller-canopied trees, thereby preventing fire from bridging into the tree canopies. Experienced ranchers in the region are also well known for successfully fighting wildfires on their own land by backfiring off their fireguards.

LOCATION OF BURN: Cattle ranch near Christoval, Texas
SIZE OF BURN: 2,460 acres
PRESCRIBED BURN APPLIED: March 12, 2008
PHOTOGRAPHS TAKEN: August 19, 2008 (5 months after burn)

Top Photo—These shin oak trees are vigorously resprouting at the base, creating available browse and diverse wildlife habitat. Bottom Photo—The blue-green grass along the road is tall dropseed, which was previously undetected on the ranch and is now increasing in abundance after the burn. Photos by George Clendenin, USDA-NRCS.

LOCATION OF BURN: Cattle ranch near Christoval, Texas
SIZE OF BURN: 2,460 acres
PRESCRIBED BURN APPLIED: March 12, 2008
PHOTOGRAPHS TAKEN: August 19, 2008 (5 months after burn)

Top Photo—This ranch has the typical shallow, rocky soils of the Edwards Plateau yet exhibits a flush of diverse new growth after a prescribed burn.
Bottom Photo—Little bluestem becomes more palatable for a time after a burn. The yucca species at the top right of the photo were top-killed, yet these resprouters will return vigorously. Photos by George Clendenin, USDA-NRCS.

LOCATION OF BURN: Livestock ranch near Christoval, Texas
SIZE OF BURN: 832 acres
PRESCRIBED BURN APPLIED: August 4, 2008
PHOTOGRAPHS TAKEN: August 15, 2008 (11 days after burn)

Top Photo—Most landowners remark that they did not realize they had so much prickly pear until they see the burned prickly pear plants after a burn. This photo shows the effectiveness of summer burning for suppressing prickly pear.
Bottom Photo—Existing ranch roads are used for firebreaks. This fire was so intense that even the prickly pear next to the road is brown, indicating high-intensity heat. Photos by George Clendenin, USDA-NRCS.

LOCATION OF BURN: Livestock ranch near Christoval, Texas
SIZE OF BURN: 271 acres
PRESCRIBED BURN APPLIED: March 20, 2009
PHOTOGRAPHS TAKEN: April 14, 2009 (25 days after burn)

Top Photo—A good fence line comparison demonstrating the vigorous new growth on the left and the old growth on the neighbor's side on the right.
Bottom Photo—Even a cool-season burn is effective in suppressing prickly pear and encouraging plant production. The oak trees in this photo are well prepared to survive future burns or wildfire, as they contain no brushy ladder fuels under their canopies. Photos by George Clendenin, USDA-NRCS.

Double Fireguard—This photo shows a 500-foot firebreak with a bladed fireguard at each end. This technique is frequently used in prescribed burns in the Concho Valley because of the highly volatile fuels and the application of large-scale burns. These double fireguards benefit landowners by providing (1) a safe firebreak free of woody canopy; (2) a method to "shut down" the burn if weather conditions deteriorate; (3) an investment for future burns; (4) grassland of high vigor for grazing animals; (5) a beneficial "edge effect" for wildlife and livestock; and (6) an area that can be back-burned to head off an oncoming wildfire. Photo by George Clendenin, USDA-NRCS.

Many prescribed burns in west central Texas are on a large scale, often encompassing several thousand acres on one burn plan in one day. The author assisted on this burn in Sterling County. Several ranchers in Sterling County have established a fire culture that they have passed on to their children. These ranchers are very experienced in planning and applying prescribed burns and in fighting off wildfires with drip torches and backfire. Photo by George Clendenin, USDA-NRCS.

# Wildlife Management in West Central Texas

Steve Nelle

6

**White-tailed deer.** Photo by Wyman Minzer.

# Managing for White-Tailed Deer in West Central Texas

White-tailed deer are an important natural resource in west central Texas. Deer hunting has become very important economically for most ranches in the area. In many cases, the income derived from deer hunting is necessary to maintain an economically viable ranching operation. The Concho Valley has become renowned in recent years for having good deer populations, good hunting opportunities, and good deer quality. Hunters from Texas and other states are discovering the quality of hunting in this area. As with any other agricultural enterprise, proper management is critical to success. Proper deer management includes the following components, which are discussed at greater length below:

### Deer Herd Management
- Deer numbers
- Sex ratio
- Age distribution of bucks
- Culling

### Deer Habitat Management
- Maintain adequate food supply for all seasons of the year
- Maintain adequate protective cover
- Maintain water availability at all times

## Deer Herd Management

Deer densities in the Concho Valley generally range from about 10 to 30 acres per deer. The deer density on any particular tract of land depends on several factors: the amount and type of brush cover, past and present livestock grazing, availability of water, and hunting practices. Factors such as supplemental feeding and high fencing also have an effect on deer numbers.

Deer numbers should be kept within the carrying capacity of the habitat. However, carrying capacity changes seasonally in response to rainfall and is very difficult to quantify. The harvest of excess deer, especially females, is one of the most important factors of deer management. Many deer managers in the Concho Valley strive to keep the deer density somewhere between 15 and 25 acres per deer. This will usually allow the habitat to remain in good condition. It will also provide good deer nutrition, which allows for good body condition, good fawn crops, and good antler development. If deer numbers are allowed to climb above the carrying capacity, key deer plants will be overgrazed and the quality of the habitat and the deer herd will decline. Progressive deer managers conduct annual deer surveys to help keep track of deer numbers and the population trend.

Most good deer managers try to keep a desirable ratio of bucks to does. Generally, managers will adjust the harvest to maintain a ratio of 1 buck for every 1 to 2 does. This may require a very aggressive doe harvest for several years and a light harvest of bucks.

Many ranchers intentionally manage their deer herds for the production of mature large-antlered bucks. A balanced mixture of young, middle-aged, and mature bucks is needed to ensure the sustained production of mature bucks in the deer herd. A balanced age structure generally consists of about one-third young bucks, one-third middle-aged bucks, and one-third mature bucks. Selective hunting practices are needed to establish and maintain this balance. Hunters should generally refrain from harvesting young and middle-aged bucks if the intent is to produce mature bucks. Bucks are considered mature when they are at least 4½ or 5½ years old. Many hunters and deer managers refrain from the harvest of trophies until bucks are 6½ years old or older for the largest antlers.

The intentional culling of bucks with less desirable antler development has become part of many deer management operations. Opinions vary widely on how the culling should be done. Mature bucks with 4 to 7 points are prime candidates for selective culling. Spikes and 3-point bucks that are 2½ years old or older should also be removed for herd improvement. The most controversial aspect of culling is the removal of yearling spikes. There is no consensus among biologists, hunters, and deer managers on the benefits of this practice, if any.

## Deer Habitat Management

Proper management of habitat is a critical component of deer management. It involves knowing the habitat requirements of deer, specifically their food, cover, and water needs. Habitat that is properly managed will generally have a more desirable population of deer, healthier deer, and better fawn crops.

Deer consume about 3.5 percent of their body weight each day

White-tailed deer habitat. Photo by Wyman Minzer.

in food. This means that an average-sized deer will eat about 1,600 pounds of food each year. Deer require a diet that is relatively high in nutritional quality, and deer are very selective feeders in order to obtain this high level of nutrition. A protein level of about 12 to 20 percent is considered desirable for deer (which is over twice as high as what is needed by cattle). Deer also require a diet high in energy and certain minerals for the best health.

The deer diet in the Concho Valley consists primarily of browse, mast, and forbs. Browse is the twigs and leaves of woody plants, such as sumac, oak, or hackberry. Mast is the fruits or seeds of woody plants, such as mesquite beans, acorns, or prickly pear fruits. Forbs are the broad-leaved herbs that many people refer to as weeds or wildflowers. Forbs provide the highest nutritional quality of all the food types, and deer prefer to eat forbs when they are available. Browse makes up the highest proportion of the diet since it is available in large quantities year-round. Evergreen browse such as juniper (cedar), prickly pear, and live oak is especially important in a dry winter. Grass makes up only 5 to 10 percent of the annual deer diet since it normally does not have the nutritional quality they need. There may be certain brief periods of the year when deer consume grass, especially winter grasses.

The best way to ensure a good food supply for deer is to maintain a large diversity of forbs, shrubs, and trees. Well-managed native rangeland in this area provides good food supplies for deer. The worst threat to the deer food supply is overgrazing by livestock and/or maintaining too many deer. Sheep and goats compete with deer. Exotic game species such as axis deer are also competitive. Serious deer managers usually limit or remove these competing animals. On land that has been heavily grazed for many years, especially by sheep and goats, restoration of desirable deer food plants may take a long time.

Cattle can also compete with deer, especially during dry periods, but the degree of competition is much less. Conservative stocking rates and rotational grazing are the best ways to ensure that there is minimal conflict between cattle and deer. During prolonged dry periods, cattle numbers should be reduced, not only to retain grass cover but also to minimize competition with deer.

Excess deer numbers also create competition for food. Proper deer herd management strives to keep the deer population at or below the carrying capacity of the habitat through hunting. In general, about 25 to 35 percent of the deer herd should be removed each year to prevent overpopulation and habitat damage. Specific harvest recommendations are based on annual deer surveys.

One of the most effective practices to enhance the deer food supply is prescribed burning. Following a fire, browse plants resprout vigorously from the base, creating a large volume of new, tender, nutritious deer forage. The newly sprouting shrubs are higher in protein, energy, and minerals and are relished by deer. There is often an increased growth of forbs after a burn since the grass cover is temporarily removed.

Food plots are sometimes used to create additional forage supplies for deer. Winter food plots usually consist of wheat or oats. Summer food plots grown for deer include cowpeas and lablab, both of which are legumes.

Deer populations thrive best where there is a large amount of protective woody cover. Deer are secretive by nature and prefer to be in close proximity to brushy cover. Deer can exist in areas of limited cover, but they prefer to have access to large blocks of dense cover. In general, ideal deer habitat will have 60 to 80 percent of the land covered by moderate to thick brush. Open areas interspersed among brushy cover are also important, since forb production is higher in these areas. Excessive brush control diminishes the quality of deer habitat, but proper brush control can enhance it. If brush is dense across large areas, it is generally desirable to conduct mechanical brush management on 20 to 40 percent of the landscape. Openings can be in the form of strips or odd-shaped clearings. They should generally be no wider than 150 to 300 feet with plenty of protective brush on each side.

For those who desire to maintain a balance between livestock production and deer habitat, a compromise between brushy cover and open areas is needed. In these situations, landowners may wish to conduct brush management on 30 to 60 percent of the landscape, which will probably lead to a reduction in deer numbers.

Fawning cover is needed to help protect newborn fawns from predators. Good fawning cover consists of areas of taller grass mixed with areas of low, dense brush. Grazing management is important in maintaining good fawning cover.

Deer need access to permanent water. If water sources dry up, deer will move to areas where they can find water. It is important to keep water maintained in all pastures, even when livestock are not present.

Within the last decade, intensive deer management in the Concho Valley has increased. This often includes high fencing and supplemental feeding. These practices greatly increase the intensity of management required as well as the cost. For most landowners, basic management as described above will yield good results and be cost-effective.

### Key for the Following Table

Deer are very selective feeders. They eat primarily forbs and browse and show a definite preference for some plants over others. Browse normally makes up the majority of the diet because of its abundance and year-round availability. Forbs are eaten when available. Grasses normally make up only 5 to 10 percent of deer diets in most seasons. This chart provides a general comparison of the forage value of the common woody plants and forbs of the region.

Class I plants are the best and most preferred forage for deer. Because these plants are highly palatable, they are the most susceptible to heavy use. On most ranches, these plants are absent or in very short supply. In some cases, they may be present in protected or inaccessible areas.

Class II plants are very desirable forage for deer. On very well-managed habitat, these plants will be common and will contribute substantially to deer diets. On overstocked or overpopulated ranges, these plants may be uncommon or may show signs of heavy use.

Class III plants are readily eaten by deer, especially if there is a shortage of Class II plants. Class III plants are often abundant and will usually provide a large percentage of the deer diet. These plants generally provide a lower level of nutrition than most Class II plants.

Class IV plants are the least preferred forage for deer and generally provide less nutritional quality than the other classes. There will usually be limited use of Class IV plants if Class II and III plants are available and abundant. During stress periods such as drought or winter, these plants may become important. If land is overstocked with goats or sheep, or overpopulated with deer, Class IV woody plants may be heavily used, which would indicate the need to reduce animal numbers.

Woody plants marked with an asterisk provide high food value for deer and livestock on a seasonal basis. Plants such as prickly pear, yucca, mesquite, and persimmon, which are often considered undesirable brush species, can be very important in producing large amounts of highly nutritious fruits, flower stalks, beans, or berries during certain seasons.

## Deer Food Plants of the Concho Valley and West Central Texas

### Woody Plants

| Class I | Class II |
|---|---|
| White honeysuckle (*Lonicera albiflora*) | Hackberry (*Celtis laevigata* var. *reticulata*) |
| Texas sophora (*Sophora affinis*) | Roemer's acacia (*Acacia roemeriana*) |
| American elm (*Ulmus americana*) | Elbowbush (*Forestiera pubescens*) |
| Mistletoe (*Phoradendron tomentosum*) | Spiny bumelia (*Sideroxylon lanuginosum*) |
| | Western soapberry (*Sapindus saponaria* var. *drummondii*) |
| | Ephedra (*Ephedra antisyphilitica*) |
| | Fourwing saltbush (*Atriplex canescens*) |
| | Old man's beard (*Clematis drummondii*) |
| | Virginia creeper (*Parthenocissus quinquefolia*) |
| | Grapevine (*Vitis* spp.) |
| | Carolina snailseed (*Cocculus carolinus*) |
| | Greenbriar (*Smilax bona-nox*) |
| | Black willow (*Salix nigra*) |

| Class III | Class IV |
|---|---|
| Live oak* (*Quercus* spp.) | Redberry juniper* (*Juniperus pinchotii*) |
| Shin oak* (*Quercus* spp.) | Ashe juniper* (*Juniperus ashei*) |
| Flameleaf sumac* (*Rhus copallinum*) | Prickly pear* (*Opuntia* spp.) |
| Littleleaf sumac* (*Rhus microphylla*) | Tasajillo* (*Opuntia leptocaulis*) |
| Evergreen sumac* (*Rhus virens*) | Yucca* (*Yucca* spp.) |
| Skunkbush sumac* (*Rhus trilobata*) | Algerita (*Berberis trifoliolata*) |
| Feather dalea (*Dalea formosa*) | Texas Persimmon* (*Diospyros texana*) |
| Buttonbush (*Cephalanthus occidentalis*) | Catclaw acacia (*Acacia greggii*) |
| Poison ivy (*Toxicodendron radicans*) | Catclaw mimosa (*Mimosa aculeaticarpa*) |
| Little walnut (*Juglans microcarpa*) | Fragrant mimosa (*Mimosa borealis*) |
| Pecan (*Carya illinoinensis*) | Honey mesquite* (*Prosopis glandulosa*) |
| Salt cedar (*Tamarix gallica*) | Wolfberry (*Lycium berlandieri*) |
| | Willow baccharis (*Baccharis neglecta*) |
| | Tickle-tongue (*Zanthoxylum hirsutum*) |
| | Lotebush (*Ziziphus obtusifolia*) |
| | Javelina bush (*Condalia ericoides*) |
| | Green condalia (*Condalia hookeri*) |
| | Whitebrush (*Aloysia gratissima*) |

*Denotes plants that provide high seasonal food value from mast, fruits, beans, or flowers.

## Perennial Forbs

| Class I | Class II |
|---|---|
| Wine cup (*Callirhoe involucrata*) | Engelmann daisy (*Engelmannia peristenia*) |
| Western primrose (*Calylophus hartwegii*) | Bush sunflower (*Simsia calva*) |
| Texas nightshade (*Solanum triquetrum*) | Velvet bundleflower (*Desmanthus velutinus*) |
| Pigeonberry (*Rivina humilis*) | Sensitive briar (*Mimosa nuttallii*) |
| Heath aster (*Aster ericoides*) | Texas snoutbean (*Rhynchosia senna*) |
| Penstemon (*Penstemon cobaea*) | Knotweed leaf-flower (*Phyllanthus polygonoides*) |
| Rain lily (*Cooperia drummondii*) | Gayfeather (*Liatris punctata*) |
| | White milkwort (*Polygala alba*) |
| | Tall goldenrod (*Solidago canadensis*) |
| | Ground cherry (*Physalis cinerascens*) |
| | Bladderpod sida (*Rhynchosida physocalyx*) |
| | Ruellia (*Ruellia metziae*) |
| | Hairy tube tongue (*Siphonoglossa pilosella*) |
| | Low menodora (*Menodora heterophylla*) |
| | Rock daisy (*Melampodium leucanthum*) |
| | Morning glory (*Ipomoea* spp.) |
| | Texas bindweed (*Convolvulus equitans*) |
| | Texas skeleton plant (*Lygodesmia texana*) |

| Class I | Class II |
| --- | --- |
| | Trailing ratany (*Krameria lanceolata*) |
| | Chalk Hill woolly-white (*Hymenopappus artemisiifolius*) |

| Class III | Class IV |
| --- | --- |
| Orange zexmenia (*Wedelia texana*) | Mealycup sage (*Salvia farinacea*) |
| Fleabane (*Erigeron strigosus*) | Texas salvia (*Salvia texana*) |
| Lazy daisy (*Aphanostephus skirrhobasis*) | Texas stillingia (*Stillingia texana*) |
| Stickleaf (*Mentzelia oligosperma*) | Rat-ear coldenia (*Tiquilia canescens*) |
| Noseburn (*Tragia ramosa*) | Threadleaf groundsel (*Senecio douglasii*) |
| Spreading sida (*Sida abutifolia*) | Grassland croton (*Croton dioicus*) |
| Indian mallow (*Abutilon fruticosum*) | Leatherweed croton (*Croton pottsii*) |
| Globemallow (*Sphaeralcea angustifolia*) | False ragweed (*Parthenium hysterophorus*) |
| Purple dalea (*Dalea lasiathera*) | Field ragweed (*Ambrosia confertiflora*) |
| Low wild mercury (*Ditaxis humilis*) | Common Dogweed (*Thymophylla pentachaeta*) |
| Mexican sagewort (*Artemisia ludoviciana*) | Gray golden aster (*Heterotheca canescens*) |
| Verbena (*Glandularia spp.*) | Plains zinnia (*Zinnia grandiflora*) |
| Cutleaf daisy (*Machaeranthera pinnatifida*) | Skullcap (*Scutellaria drummondii*) |
| Lindheimer's Copperleaf (*Acalypha phleoides*) | Milkweeds (*Asclepias*) |
| Western ragweed (*Ambrosia psilostachya*) | Wavy-leaf milkvine (*Funastrum crispum*) |
| | Broom snakeweed (*Gutierrezia sarothrae*) |
| | Twin-leaf senna (*Senna roemeriana*) |
| | Silverleaf nightshade (*Solanum elaeagnifolium*) |
| | Western horsenettle (*Solanum dimidiatum*) |
| | Upright Prairie coneflower (*Ratibida columnifera*) |
| | Curlycup gumweed (*Grindelia nuda*) |
| | Dutchman's breeches (*Thamnosma texana*) |

## Managing for Turkeys in West Central Texas

West central Texas supports good populations of Rio Grande turkeys. The presence of turkeys on any given piece of land depends on the kind of habitat provided and its proximity to other habitats. Turkeys range for long distances and cover a large area in their annual life cycle. It is not uncommon for a group of turkeys to travel 10 to 15 miles from their fall and winter roosting areas to their spring and summer ranges. For this reason, a ranch may have turkeys for only a portion of the year, although some places will have them all year

Rio Grande turkey. Photo by Wyman Minzer.

long. The number of turkeys on a piece of property may fluctuate dramatically from season to season and year to year.

In the fall and winter, turkeys tend to migrate to river and creek bottoms, where they congregate in large groves of tall trees for roosting. These winter roosting areas may sometimes support very large concentrations of turkeys. Landowners who desire to maintain turkey habitat are careful not to disturb these key roosting areas. Turkeys are very adaptable; if creek bottoms are not available for roosting, they will sometimes roost on power poles, tank batteries, or pump jacks.

Near the end of winter, turkeys begin to disperse to their nesting grounds. They break up into smaller groups and travel to the uplands and hills where they will spend the spring and summer and possibly raise a crop of young turkeys. Turkey hens usually prefer to nest in vegetation at least 18 inches tall. Nest locations include tall grass, tall weeds, dense vines, or brush. Nests are generally located within one-quarter mile of water. The majority of turkey nests are usually destroyed by predators, including raccoons, skunks, foxes, bobcats, coyotes, and snakes. The presence of good nesting cover over a large area is the best way to minimize nest predation. Predator control has not been shown to increase turkey numbers. Hens will renest if their first nest is destroyed, but subsequent nests are seldom successful.

Most turkey eggs hatch from mid-May to mid-June, with an average brood of 8 to 10. Young poults eat primarily insects. If a good crop of insects is not available, the survival of young turkeys will be poor. For this reason, a dry spring usually results in a poor hatch or poor survival of young turkeys. A wet spring with lots of insects often results in a good hatch and good turkey survival.

Adult turkeys consume a very wide variety of food items. They eat insects whenever available, but otherwise they eat a variety of berries, other fruits, and seeds of trees, shrubs, vines, forbs, and grasses. They also consume large quantities of green foliage from grasses and forbs and make heavy use of agricultural crops such as wheat and grain sorghum.

Turkeys drink water on a daily basis. Landowners who want to maintain turkey populations should be careful to maintain year-round water availability in all pastures. Conventional livestock watering troughs are suitable for turkeys, but escape ramps should be placed in troughs in case young poults or other wildlife fall into the water. A better way to provide water for turkeys and other wildlife is to create ground-level earthen depressions that are fed from pipelines or overflow from windmills and water storage tanks. These depressions provide not only water but also a fringe of green vegetation, which in turn produces insects and seeds.

The best management for turkeys is to retain large areas of diverse native vegetation and habitat types including a combination of brushy cover and open grassy areas. Food plots using wheat or grain sorghum can be planted to attract turkeys.

Turkeys can also be attracted by supplemental feeding. Grains such as corn or grain sorghum are commonly used to attract turkeys. Protein supplements intended for deer or livestock are also consumed by turkeys. Landowners should be aware that turkeys are susceptible to aflatoxin, which is sometimes found in grains and grain products. Those who wish to feed turkeys need to ensure that the feed has been tested and is safe for them and that it is stored in a dry place.

## Turkey Food Plants of the Concho Valley and West-Central Texas

| Grasses | Woody Plants | Forbs |
| --- | --- | --- |
| Slim tridens (*Tridens muticus* var. *muticus*) | Spiny bumelia (*Sideroxylon lanuginosum*) | Wild onion (*Allium drummondii*) |
| White tridens (*Tridens albescens*) | Prickly pear (*Opuntia* spp.) | Peavine (*Astragalus nuttallianus*) |
| Plains bristlegrass (*Setaria leucopila*) | Tasajillo (*Opuntia leptocaulis*) | Deer pea vetch (*Vicia ludoviciana*) |
| Reverchon bristlegrass (*Setaria reverchonii*) | Texas persimmon (*Diospyros texana*) | Filaree (*Erodium* spp.) |
| Texas cupgrass (*Eriochloa sericea*) | Pecan (*Carya illinoinensis*) | Plantain (*Plantago* spp.) |
| Switchgrass (*Panicum virgatum*) | Little walnut (*Juglans microcarpa*) | Bur clover (*Medicago minima*) |
| Yellow Indiangrass (*Sorghastrum nutans*) | Wild plum (*Prunus americana*) | Pigweed (*Amaranthus* spp.) |
| Vine-mesquite (*Panicum obtusum*) | Western soapberry (*Sapindus saponaria* var. *drummondii*) | Ground cherry (*Physalis cinerascens*) |
| Sideoats grama (*Bouteloua curtipendula*) | Littleleaf sumac (*Rhus microphylla*) | Texas nightshade (*Solanum triquetrum*) |
| Sand dropseed (*Sporobolus cryptandrus*) | Skunkbush sumac (*Rhus trilobata*) | Silverleaf nightshade (*Solanum elaeagnifolium*) |
| Little barley (*Critesion pusillum*) | Poison ivy (*Toxicodendron radicans*) | Low wild mercury (*Ditaxis humilis*) |
| Tobosagrass (*Pleuraphis mutica*) | Live oak (*Quercus virginiana*) acorns | Indian mallow (*Abutilon fruticosum*) |
| Flat sedge (*Cyperus* spp.) | Honey mesquite (*Prosopis glandulosa*) beans | Pigeonberry (*Rivina humilis*) |
| Cedar sedge (*Carex planostachys*) | Juniper (*Juniperus* spp.) berries | Gaura (*Gaura* spp.) |
| Texas wintergrass (*Nassella leucotricha*) | Lotebush (*Ziziphus obtusifolia*) berries | Primrose (*Calylophus* and *Oenothera*) |
| Johnsongrass (*Sorghum halepense*) | Algerita (*Berberis trifoliolata*) | Croton (*Croton* spp.) |
| Squirreltail (*Elymus longifolius*) | Ephedra (*Ephedra antisyphilitica*) | Noseburn (*Tragia ramosa*) |
| Hooded windmillgrass (*Chloris cucullata*) | Wolfberry (*Lycium berlandieri*) | Spiderwort (*Tradescantia* spp.) |
| Wildrye (*Elymus* spp.) | Catclaw acacia (*Acacia greggii*) | Rain lily (*Cooperia drummondii*) |
| | Mexican buckeye (*Ungnadia speciosa*) | |
| | Texas mulberry (*Morus microphylla*) | |
| | Red mulberry (*Morus rubra*) | |
| | Carolina snailseed (*Cocculus carolinus*) | |
| | Grapevine (*Vitis* spp.) | |
| | Virginia creeper (*Parthenocissus quinquefolia*) | |

The normal lifespan for turkeys is about 3 years. For this reason, a good hatch is not needed each year to perpetuate populations. A good hatch every other year is sufficient to maintain numbers. Turkey numbers will sharply decline if there are 2 successive years of poor reproduction.

# Managing for Doves in West Central Texas

Traditionally, most of the interest in dove hunting in west central Texas has been for mourning doves. In recent years, the increase in white-winged doves has added another dimension to dove hunting and dove management.

Dove management revolves around two primary things: food and water. Landowners who wish to attract doves will generally manage food and/or water to provide the best hunting opportunities.

Doves are almost exclusively seed eaters. They make heavy use of agricultural grains (wheat and grain sorghum), as well as the seeds of native plants. Farmers who wish to attract doves to their property can do several things to provide access to waste grain after harvest. Delaying the plowing of fields for several months will leave waste grain available on the soil surface. Once the field is plowed, access to grain is greatly reduced. Leaving some areas of grain unharvested can provide large numbers of seeds. Some farmers will leave several rows of standing grain unharvested around the outer edge of fields where dove hunting will occur. Other crops that can be used to feed doves include commercial sunflowers and sesame. Grain and seed areas are often manipulated by mowing or dragging part of the field prior to dove season and are staggered periodically throughout the season. This manipulation scatters the seeds on the ground, where doves can readily find them. Doves are reluctant to land and search for seeds in heavily vegetated areas. They prefer sparse vegetation and lots of bare ground where they can see approaching danger.

Doves also eat many types of native seeds. These include native sunflower, pricklypoppy, croton, buffalobur, spurges, pigweed, Johnsongrass, and many others. Some of these plants grow on native rangeland pastures and others grow best in disturbed areas specifically managed for doves. Those who wish to grow native common sunflower for doves should plant seeds in the fall or early winter since the seeds need a prolonged period of cold to break dormancy. A worthwhile food plot for doves consists of a combination of wheat and sunflower.

Landowners and dove hunters should be aware of federal baiting laws (http://www.fws.gov/le/what-is-legal.html). Intentional baiting of doves by distributing seeds or grain is prohibited. This may also apply to unintentional baiting where doves have been attracted to deer or quail feeding stations. The best advice is to check with local game wardens to ensure a proper understanding of baiting laws.

Mourning dove. Photo by Wyman Minzer.

Doves drink water every day, and landowners who provide the right kind of water can sometimes attract large numbers of them. Doves will drink from puddles, creeks, ponds, lakes, and even water troughs. They prefer to drink in areas that are devoid of vegetation, as they want to be able to see in all directions while drinking. Water holes surrounded by bare ground will attract more doves than water holes lined with vegetation. Doves are often observed perching in a nearby tree prior to flying down to water. The presence of trees, especially dead trees near water holes, seems to entice greater dove use. Providing water is a more effective dove hunting strategy where other water sources are scarce.

White-winged doves reside mostly in cities and towns, where they roost in neighborhoods with larger trees. These doves frequently fly out to rural areas, especially farmland, to feed.

The small Inca dove is also common in the region but is not a game species and cannot be legally hunted. Many homeowners who feed birds find it easy to attract Inca doves to their yards.

A fourth species of dove is becoming more and more common in the Concho Valley, especially near towns. The Eurasian collared dove is not native but has become abundant in some areas. The ecological impact of this new exotic species on native doves is not known. This species may be legally hunted during dove season.

## Managing for Quail in West Central Texas

West central Texas is home to two species of native quail. Bobwhite quail are the most common species, especially in the eastern and central part of the region. Scaled quail (blue quail) can be found across the region but are more common in the western part of the Concho Valley. Bobwhites tend to thrive in areas with better grass cover, while scaled quail are often found in sparser cover. Many ranches have a combination of the two species, but the proportion will vary from year to year.

Quail numbers fluctuate drastically, and their numbers range from almost none in some years to abundant in others. As a general rule, the best quail years are the wet years, and the worst times are during and following dry years. Because of their dependence on favorable rainfall, quail are a challenge to manage, and to a large degree this can depend on factors beyond the control of landowners. Nevertheless, landowners who understand the habitat requirements of quail

**Bobwhite quail. Photo by Wyman Minzer.**

Scaled quail (blue quail). Photo by Wyman Minzer.

can manage their land for the best chance of maintaining quail populations.

Quail need a specific combination of habitat types, ideally in close proximity to each other. Quail nest at the base of large grass clumps with a large proportion of dry leaves and stems left from the previous year. The ideal nest site is between two or three overlapping grass clumps. The best way to visualize good nesting cover is to think of grass clumps about the size of a basketball, or slightly smaller. These large grass clumps should be abundant and spaced no more than about 10 to 15 feet apart. Grasses that create good nesting cover include little bluestem, tobosagrass, vine-mesquite, sideoats grama, and other grasses of similar stature. Pastures that are substantially grazed generally lack ideal nesting cover. Areas that lack adequate nesting cover are more vulnerable to nest predation from raccoons, skunks, foxes, bobcats, coyotes, and snakes. Light grazing, rotational grazing, and seasonal grazing are techniques that can help ensure adequate grass cover for quail nesting.

Research in the Concho Valley has shown that quail will also use prickly pear clumps for nest concealment, especially in the absence of good grass cover. For this reason, ranchers may want to consider retaining some areas of moderate prickly pear communities.

Interspersed among good grass cover, quail also need a good distribution of low-growing shrubs for protection from hawks. Low, dense brush such as littleleaf sumac, lotebush, and algerita can provide this kind of cover. These escape coverts, or "quail houses" (sometimes called loafing cover), should be open beneath and dense above, and they should be present about every 50 to 100 feet in good quail habitat. This kind of brushy cover should be specifically retained during brush-control operations.

The quail diet consists primarily of seeds, with large seeds being more valuable than small seeds. Quail utilize the seeds of forbs, woody plants, and grasses. Some of the primary seeds used by quail in the Concho Valley include those of croton, spurges, sunflower, prickly-poppy, buffalobur, tickle tongue, bumelia, and mesquite.

Quail will eat insects anytime they are available. Young chicks are especially dependent on bugs for their primary food source during the first 6 weeks of life. A dry spring without a good crop of insects is likely to be a bad quail year. In the late winter, quail will often eat greens. These are the small tender leaves of weeds and forbs such as filaree, plantain, bur clover, and other winter weeds.

There is considerable debate on the issue of quail and water. Quail need water, just like any other animal, but they can usually get most if not all of their water from insects, berries, dew, and greens. Quail can even manufacture water from a chemical reaction that takes place during the digestion of carbohydrates. The debate revolves around the value of standing water for quail. Quail will readily drink from puddles, troughs, ponds, or other water sources, but such use does not necessarily indicate that they require standing water. Some ranchers believe they have improved habitat and increased quail numbers by the installation of water. In the western, more arid part of the Concho Valley, water development may be of value, especially in dry years. Traditional livestock water troughs can be made more useful for quail by setting the float to allow the water to come right up near the lip of the trough and by installing escape ramps so that small quail that fall into the water can get out. Ground-level water is more useful to many species of wildlife than water troughs.

In summary, quail need a combination of grassy cover for nesting,

## Quail Food Plants of the Concho Valley and West-Central Texas

| Grasses and Grasslike Plants (Seeds) | Woody Plants (Berries / Fruits) | Forbs (Seeds) |
|---|---|---|
| Slim tridens (*Tridens muticus* var. *muticus*) | Netleaf hackberry (*Celtis laevigata* var. *reticulata*) | Grassland croton (*Croton dioicus*) |
| Plains bristlegrass (*Setaria leucopila*) | Spiny bumelia (*Sideroxylon lanuginosum*) | Leatherweed croton (*Croton pottsii*) |
| Reverchon bristlegrass (*Setaria reverchonii*) | Wolfberry (*Lycium berlandieri*) | One-seed croton (*Croton monanthogynus*) |
| Texas cupgrass (*Eriochloa sericea*) | Honey mesquite (*Prosopis glandulosa*) beans | Prostrate spurge (*Euphorbia* spp.) |
| Hall panicum (*Panicum hallii*) | Tickle tongue (*Zanthoxylum hirsutum*) | Hoary euphorbia (*Chamaesyce lata*) |
| Browntop panicum (*Urochloa ramosa*) | Littleleaf sumac (*Rhus microphylla*) | Snow-on-the-mountain (*Euphorbia marginata*) |
| Vine-mesquite (*Panicum obtusum*) | Skunkbush sumac (*Rhus trilobata*) | Toothed spurge (*Euphorbia dentata*) |
| Sideoats grama (*Bouteloua curtipendula*) | Flameleaf sumac (*Rhus copallinum*) | Bundleflower (*Desmanthus* spp.) |
| Sand dropseed (*Sporobolus cryptandrus*) | Prickly pear (*Opuntia* spp.) | Sensitive briar (*Mimosa nuttallii*) |
| Cedar sedge (*Carex planostachys*) | Tasajillo (*Opuntia leptocaulis*) | Broom snakeweed (*Gutierrezia sarothrae*) |
| Johnsongrass (*Sorghum halepense*) | Oak (*Quercus* spp.) acorns | Common broomweed (*Gutierrezia dracunculoides*) |
| Columbus grass (*Sorghum almum*) | Juniper (*Juniperus* spp.) berries | Cowpen daisy (*Verbesina encelioides*) |
| Rescue grass (*Bromus catharticus*) | Lotebush (*Ziziphus obtusifolia*) berries | Common sunflower (*Helianthus annuus*) |
| | Green condalia (*Condalia hookeri*) | Buffalobur (*Solanum rostratum*) |
| | Algerita (*Berberis trifoliolata*) | Pigweed (*Amaranthus* spp.) |
| | Ephedra (*Ephedra antisyphilitica*) | White pricklypoppy (*Argemone aurantiaca*) |
| | Catclaw acacia (*Acacia greggii*) | Basketflower (*Centaurea americana*) |
| | Roemer's acacia (*Acacia roemeriana*) | Nuttall peavine (*Astragalus nuttallianus*) |
| | Fragrant mimosa (*Mimosa borealis*) | Deer pea vetch (*Vicia ludoviciana*) |
| | Catclaw mimosa (*Mimosa aculeaticarpa*) | Lambsquarter (*Chenopodium album*) |
| | Carolina snailseed (*Cocculus carolinus*) | Low menodora (*Menodora heterophylla*) |
| | Texas nightshade (*Solanum triquetrum*) | Texas stillingia (*Stillingia texana*) |
| | | Noseburn (*Tragia ramosa*) |
| | | Low wild mercury (*Ditaxis humilis*) |
| | | Globemallow (*Sphaeralcea* spp.) |
| | | Indian mallow (*Abutilon fruticosum*) |
| | | Spreading sida (*Sida abutifolia*) |
| | | Dutchman's breeches (*Thamnosma texana*) |
| | | Rat-ear coldenia (*Tiquilia canescens*) |
| | | Ground cherry (*Physalis cinerascens*) |

woody cover for escape, and weeds and forbs for seed production. Conservative grazing and carefully planned brush control are the primary tools used by quail managers. In years of favorable and well distributed rainfall, quail will be able to take advantage of good habitat. In very dry years, there may be little or no quail production even where habitat is excellent.

Since quail usually spend their entire life in a rather small area, they must have access to nesting cover, escape cover, and a year-round food source in close proximity to each other. In order to create the most habitable area for quail, these habitat elements should all be present within each 40- to 80-acre area.

## Birds of West Central Texas and Their Management

West central Texas is home to a large variety and abundance of birds, as well as people who are interested in birds. Traditionally, much emphasis has been placed on the hunting of game birds, especially doves, quail, and turkeys. But more recently a larger interest has developed around bird-watching and the conservation of birds. Bird-watchers describe their hobby as "birding" and refer to themselves as "birders." The Concho Valley has a large and growing cadre of committed and skilled birders. Birders are often some of the best naturalists and have knowledge about native plants, soils, ecology, and land management.

Many landowners and homeowners in the Concho Valley are interested in attracting and maintaining a diverse bird population on their land and in their backyards. Much of this interest is for personal enjoyment, but there is also a growing interest in ecotourism, which may bring birders from long distances to enjoy the bird life of this region.

The birds of the Concho Valley can be divided into several general categories. Raptors and large birds of prey include hawks, eagles, kites, falcons, owls, and buzzards. Water birds include ducks, geese, herons, egrets, pelicans, grebes, cranes, coots, plovers, sandpipers, kingfishers, gulls, and terns. Songbirds include myriad smaller birds such as flycatchers, jays, chickadees, wrens, thrashers, thrushes, bluebirds, gnatcatchers, waxwings, shrikes, vireos, warblers, orioles, ta-

Female northern cardinal. Photo by Beverly Moseley, USDA-NRCS.

nagers, finches, juncos, sparrows, buntings, grosbeaks, and towhees. Other bird groups that are often combined with the songbirds include hummingbirds, swifts, woodpeckers, nighthawks, and cuckoos. Within each of these subgroups are often several to many species.

Those who wish to attract birds to their property or manage their land for the benefit of birds need a basic knowledge of bird habitat requirements. Each species of bird has its own specific requirements for food, cover, and water. It would be difficult if not impossible to integrate the specific habitat requirements of hundreds of different bird species into a single habitat management plan. A more practical approach is to manage lands for the greatest habitat diversity possible. In this way, a landowner or homeowner has the greatest opportunity to provide favorable conditions for a large number of bird species. Obviously, the homeowner or suburban resident with a small plot of land will not have as much opportunity for this as a rancher with hundreds or perhaps thousands of acres of native habitat.

**Basic Habitat for Birds: Food, Cover, and Water**

Songbirds eat a large variety of food items including insects and other invertebrates, seeds, and fruits. Some songbirds consume only seeds, others only insects, and some a combination of food types. The shape of the beak is the best indication of the type of food eaten. Stout, thick beaks (such as those of cardinals) usually indicate a preference for seeds. Thin, short, pointed beaks (such as those of warblers) indicate a preference for small insects. Birds with medium-length beaks often eat a variety of insects, seeds, and fruits. The beaks of hummingbirds enable them to sip nectar from long, tubular flowers. Woodpecker beaks are designed to bore into tree bark in search of insects.

The largest variety of plants, including grasses, forbs, vines, shrubs, and trees, will always provide the most suitable food supply for the largest variety of songbirds. This same plant variety will also support a large number and diversity of insects. Below is a partial listing of Concho Valley plants that provide seeds and/or fruits eaten by songbirds.

- Shrubs—elbowbush (*Forestiera pubescens*), sumacs (*Rhus* spp.), tickle tongue (*Zanthoxylum hirsutum*), condalia (*Condalia hookeri*), lotebush (*Ziziphus obtusifolia*), algerita (*Berberis trifoliolata*), prickly pear (*Opuntia* spp.), tasajillo (*Opuntia leptocaulis*), wolfberry (*Lycium berlandieri*), honeysuckle (*Lonicera albiflora*), acacia (*Acacia* spp.), mimosa (*Mimosa* spp.), yucca (*Yucca* spp.)
- Trees—oaks (*Quercus* spp.), pecan (*Carya illinoinensis*), mulberry (*Morus rubra*), bumelia (*Sideroxylon lanuginosum*), hackberry (*Celtis* spp.), elms (*Ulmus* spp.), juniper (*Juniperus* spp.), mesquite (*Prosopis glandulosa*), and the tree parasite, mistletoe (*Phoradendron tomentosum*)
- Vines—Virginia creeper (*Parthenocissus quinquefolia*), grapes (*Vitis* spp.), Carolina snailseed (*Cocculus carolinus*), greenbriar (*Smilax bona-nox*), poison ivy (*Toxicodendron radicans*), ivy treebine (*Cissus incisa*), dewberry (*Rubus* spp.)
- Forbs—ragweed (*Ambrosia* spp.), pigweed (*Amaranthus* spp.), various spurge species, woodsorrel (*Oxalis* spp.), low wild

mercury (*Ditaxis humilis*), various croton species, noseburn (*Tragia* spp.), nightshade (*Solanum* spp.), ground cherry (*Physalis cinerascens*), various aster species, goldenrod (*Solidago canadensis*), purple eryngo (*Eryngium leavenworthii*), sunflower (*Helianthus* spp.), Engelmann daisy (*Engelmannia peristenia*), gayfeather (*Liatris* spp.), cowpen daisy (*Verbesina encelioides*), plantain (*Plantago* spp.), filaree (*Erodium* spp.), various gaillardia species, lambsquarter (*Chenopodium album*), ruellia (*Ruellia metziae*), verbena (*Glandularia* spp.), pigeonberrry (*Rivina humilis*), knotweed leaf-flower (*Phyllanthus polygonoides*), wine cup (*Callirhoe involucrata*), various sida species, Indian mallow (*Abutilon fruticosum*), bundleflower (*Desmanthus* spp.), various peavine and vetch species, various dalea species, bluebonnet (*Lupinus subcarnosus*), pepperweed (*Lepidium virginicum*), coldenia (*Tiquilia canescens*), broomweed (*Gutierrezia dracunculoides*), various coneflower species, wild onion (*Allium drummondii*), bladderpod (*Rhynchosida physocalyx*), menodora (*Menodora heterophylla*), pricklypoppy (*Argemone aurantiaca*)
- Grasses and grasslike—bristlegrass (*Setaria* spp.), various panicum species, barnyard grass (*Echinochloa crus-galli*), dropseed (*Sporobolus* spp.), various paspalum species, various grama grasses, tridens (*Tridens* spp.), signal grass (*Urochloa platyphylla*), various sedges

**Cover**

In order to meet the cover requirements for a variety of songbirds, a diversity of cover types is needed. Cover for nest concealment is important, and each species of bird has its own unique preferences for nest placement. Cover is also needed for fledging of young, searching for food, roosting, and protection from the elements and from predators. Some of the major cover types used by Concho Valley birds include the following:

- Grassland dominated by dense midgrasses and perennial forbs
- Grassland dominated by sparse midgrasses and short grasses with significant bare ground and scattered shrubs
- Mixed shrubland interspersed with grassland
- Dense, low-growing shrubs

- Oak savanna interspersed with grassland
- Mixed juniper/oak woodland or shrubland
- Multilayered riparian woodland with a dense understory of shorter trees, shrubs, and vines

The ideal habitat for a variety of birds consists of a mosaic of several of these cover types, preferably close to each other and to water.

### Water

Songbirds get much of their water from insects or from fleshy fruits and berries. Natural and human-made sources of surface water such as creeks, ponds, puddles, troughs, windmill overflows, and birdbaths are also used by many kinds of songbirds when available.

Traditional water development for livestock can be modified to be more useful to birds. Ramps can be installed, large rocks can be stacked, or floating platforms can be placed in water troughs to provide access and escape. Float valves or standpipes in storage tanks or troughs can be adjusted to allow a slight overflow. This overflow can be piped to an earthen depression or merely allowed to form small puddles and pools. Pipes from windmills can be modified to create drips that will be used by birds.

Taps and valves can be installed in existing pipelines to provide additional ground-level water. Small birdbath-shaped depressions can be formed with rock and mortar, concrete, fiberglass, or plastic liners. Small valves or drip emitters can be used to regulate the flow.

Ponds can be constructed if soil and topography allow, but suitable potential pond locations are very rare in this region of Texas because of soil limitations. Ponds that are partially fenced from livestock provide better cover and more natural conditions for many birds. Most ponds in the Concho Valley provide only temporary water and are dry during critical periods.

"Guzzlers" can be constructed where other traditional water development is not feasible. These include the collection of rainwater from a catchment apron to a storage tank, where it is then delivered to a small drinker at ground level.

### Bird Habitat Management Techniques

The following habitat management techniques do not all apply equally to all properties, but an appropriate combination of these techniques can be used to enhance food and cover.

1. Enhance the establishment and growth of desirable trees, shrubs, vines, and forbs by removing or reducing goats, sheep, and exotics and keeping deer numbers low.
2. Leave dense wooded areas along creeks, canyons, and riparian areas.
3. Retain existing groves, thickets, and mottes of taller trees, especially where these are less common.
4. Where juniper, mesquite, or prickly pear has become overabundant and where a more open landscape is desired, remove them individually or thin them mechanically or chemically to the desired density. Leave desirable shrubs intact. It is advisable to remove brush in small patches and increments instead of conducting large-scale operations.
5. Where juniper is encroaching into or near the canopy of more desirable trees or shrubs, selectively remove or reduce the juniper.
6. Following brush control, leave dead slash on the ground; this favors the establishment and growth of desirable grasses, forbs, shrubs, and trees. The slash provides protection, mulch, and shade. Birds that perch on the dead branches often deposit seeds.
7. The microhabitat created under the canopy of larger mesquite is favorable for the growth of a different plant community than would otherwise be found. These areas are shadier and cooler and have enriched soil. Consider retaining numerous scattered large mesquites when conducting brush management.
8. Areas that have not been grazed for many years may develop an excessively thick grass cover. Use a planned grazing system to reduce the grass cover and improve plant diversity. Combinations of mowing and/or burning can also be used.
9. Areas that have been heavily grazed will benefit from several years of no grazing.
10. Light and periodic grazing favors better plant diversity than moderate, heavy, or continuous grazing.
11. Prescribed burning can increase the diversity and abundance of desirable forbs but will temporarily reduce the canopy and fruit production of vines, shrubs, and trees. Numerous smaller burns interrupted by unburned areas are better than larger continuous burns. To avoid damage to nesting cover

during the nesting season, conduct burning in the cool season prior to the middle of March.

12. Burning under cooler conditions will allow fires to creep under trees without damage to the canopy. Burning under harsher conditions will damage shrub and tree canopies, which may take several years to recover.
13. If juniper, mesquite, or other woody plants are removed mechanically, seed the disturbed areas with desirable native forbs and grasses.
14. Where a greater variety and abundance of seeds is desired for ground-feeding birds, disking or tilling of strips will stimulate the growth of a variety of large-seeded forbs such as pricklypoppy, spurge, snow-on-the-mountain, croton, peavine, and sunflower.
15. Where the needed diversity of shrubs or trees is not present, planting can be done. Many native plant nurseries carry a good variety of containerized shrubs and trees. These need to be adapted to the site, watered regularly the first two or three years, and protected from deer and livestock. This is usually feasible only in small areas.
16. Where water is available, the periodic irrigation of small plots of native habitat can provide a rich source of succulent vegetation and insects. Sprinklers set along a pipeline to keep small areas green can help ensure insect production even in very dry periods.
17. Use insecticides with great care and restraint since insects are a very important food for songbirds.
18. Snags (dead standing trees and branches) are an important feature for many songbirds. They are used for nest cavities and perches. Maintain dead standing trees. Where more snags are desired, kill selected trees with a diameter of at least 6 inches using approved herbicide treatments, diesel oil, or mechanical girdling.
19. Feeding of birds will attract and concentrate birds for viewing but will probably not help increase bird populations. Feeding with a large variety of seeds, fruits, nuts, and suet and using different feeder types in a variety of locations near cover and water will be most effective.

Western scrub jay. Photo by Beverly Moseley, USDA-NRCS.

## Improving Songbird Reproduction

Nest parasitism by the brown-headed cowbird has been shown to seriously reduce successful reproduction of many species of songbird. Cowbirds lay their eggs in the nests of other birds and often destroy the eggs and nestlings of the smaller songbirds. The cowbirds abandon their own eggs and allow the songbird to incubate and raise the young cowbirds. Trapping of cowbirds from March through May has proven to reduce nest parasitism and improve songbird nesting success. Trapping must be done under a permit from the Texas Parks and Wildlife Department. Several cowbird trapping efforts have been carried out in the Concho Valley and have resulted in substantial improvements in songbird populations.

## Native and Regionally Adapted Plants for Hummingbirds and Butterflies

| Hummingbirds | Butterfly Nectar |
|---|---|
| **WOODY PLANTS** | **WOODY PLANTS** |
| *Salvia greggii* | *Salvia greggii* |
| Texas lantana (*Lantana horrida*) | Acanthus (*Anisacanthus* spp.) |
| Acanthus (*Anisacanthus* spp.) | Texas lantana (*Lantana horrida*) |
| White honeysuckle (*Lonicera albiflora*) | Baccharis (*Baccharis* spp.) |
| Desert willow (*Chilopsis linearis*) | Wild plum (*Prunus americana*) |
| Trumpet vine (*Campsis radicans*) | Hawthorn (*Crataegus* spp.) |
| Yellow bells (*Tecoma stans*) | Ceanothus (*Ceanothus* spp.) |
| Redbud (*Cercis Canadensis*) | Sumac (*Sumac* spp.) |
| False indigo (*Amorpha fruticosa*) | Elbowbush (*Forestiera pubescens*) |
| Agave (*Agave* spp.) | Buttonbush (*Cephalanthus occidentalis*) |
| Red yucca (*Hesperaloe parviflora*) | Whitebrush (*Aloysia gratissima*) |
| Ocotillo (*Fouquieria splendens*) | Kidneywood (*Eysenhardtia texana* |
| | White honeysuckle (*Lonicera albiflora*) |
| **PERENNIAL FORBS** | |
| Salvia (all) (*Salvia* spp.) | **PERENNIAL FORBS** |
| Turk's cap (*Malvaviscus arboreus*) | Mistflower (*Conoclinium coelestinum*) |
| Penstemon (all) (*Penstemon* spp.) | Heath aster (*Aster ericoides*) |
| Rock betony (*Stachys coccinea*) | Salvia (*Salvia* spp.) |
| Columbine (*Aquilegia canadensis*) | Gayfeather (*Liatris punctata*) |
| Cardinal flower (*Lobelia cardinalis*) | Ironweed (*Vernonia* spp.) |
| Texas paintbrush (*Castilleja indivisa*) | Rock daisy (*Melampodium leucanthum*) |
| **ANNUAL / BIENNIAL FORBS** | Chocolate daisy (*Berlandiera lyrata*) |
| Standing cypress (*Ipomopsis rubra*) | Engelmann daisy (*Engelmannia peristenia*) |
| Texas bluebonnet (*Lupinus subcarnosus*) | Orange zexmenia (*Wedelia texana*) |
| Phlox (*Phlox* spp.) | Bush sunflower (*Simsia calva*) |
| Lemon beebalm (*Monarda citriodora*) | Upright prairie coneflower (*Ratibida columnifera*) |
| Texas paintbrush (*Castilleja indivisa*) | Maximilian sunflower (*Helianthus maximiliani*) |
| | Goldenrod (*Solidago* spp.) |
| | Milkweed (*Asclepias* spp.) |
| | Verbena (*Glandularia* spp.) |
| | Texas frogfruit (*Phyla nodiflora*) |
| | Penstemon (*Penstemon cobaea*) |
| | Pavonia (*Pavonia* spp.) |
| | Prairie clover (*Dalea purpurea*) |
| | Turk's cap (*Malvaviscus arboreus*) |
| | |
| | **ANNUAL / BIENNIAL FORBS** |
| | White pricklypoppy (*Argemone aurantiaca*) |
| | Coreopsis (*Coreopsis basalis*) |
| | Huisache daisy (*Amblyolepis setigera*) |
| | Indian blanket (*Gaillardia pulchella*) |
| | Cowpen daisy (*Verbesina encelioides*) |
| | Lemon beebalm (*Monarda citriodora*) |
| | Standing cypress (*Ipomopsis rubra*) |

Butterfly Larvae Food

WOODY PLANTS
Carolina buckthorn (*Frangula caroliniana*)
False indigo (*Amorpha fruticosa*)
Acanthus (*Anisacanthus* spp.)
Sumac (*Sumac* spp.)
Wild plum (*Prunus americana*)
Kidneywood (*Eysenhardtia texana*)
Texas lantana (*Lantana horrida*)
Tickle tongue (*Zanthoxylum hirsutum*)
Western soapberry (*Sapindus saponaria* var. *drummondii*)
Hackberry (*Celtis* spp.)
Oaks (*Quercus* spp.)
Old man's beard (*Clematis drummondii*)
Juniper (*Juniperus* spp.)
Mesquite (*Prosopis glandulosa*)
Mexican buckeye (*Ungnadia speciosa*)
Wild plum (*Prunus americana*)
Fourwing saltbush (*Atriplex canescens*)
Sacahuista (*Nolina texana*)

FORBS
Greenthread (*Thelesperma filifolium*)
Sagewort (*Artemisia ludoviciana*)
Heath aster (*Aster ericoides*)
Texas paintbrush (*Castilleja indivisa*)
Globemallow (*Sphaeralcea* spp.)
Wine cup (*Callirhoe involucrate*)
Hairy tube tongue (*Siphonoglossa pilosella*)
Flax (*Linum* spp.)
Dalea (*Dalea* spp.)
Twin-leaf senna (*Callirhoe involucrata*)
Mistletoe (*Phoradendron tomentosum*)
Pigeonberry (*Rivina humilis*)
Passionflower (*Passiflora* spp.)
Milkweed (*Asclepias* spp.)
Ruellia (*Ruellia metziae*)
Dutchman's pipe (*Aristolochia coryi*)
Noseburn (*Tragia ramosa*)
Croton (*Croton* spp.)
Prairie acacia (*Acacia angustissima*)

GRASSES
Yellow Indiangrass (*Sorghastrum nutans*)
Big bluestem (*Andropogon gerardii*)
Switchgrass (*Panicum virgatum*)
Little bluestem (*Schizachyrium scoparium*)
Sideoats grama (*Bouteloua curtipendula*)
Blue grama (*Chondrosum gracile*)
Hairy tridens (*Erioneuron pilosum*)
Dichanthelium (*Dichanthelium* spp.)
Cedar sedge (*Carex planostachys*)

Monarch butterfly on a prairie coneflower. Photo by USDA-NRCS.

# Appendix A. Leaf Shapes and Leaf Types

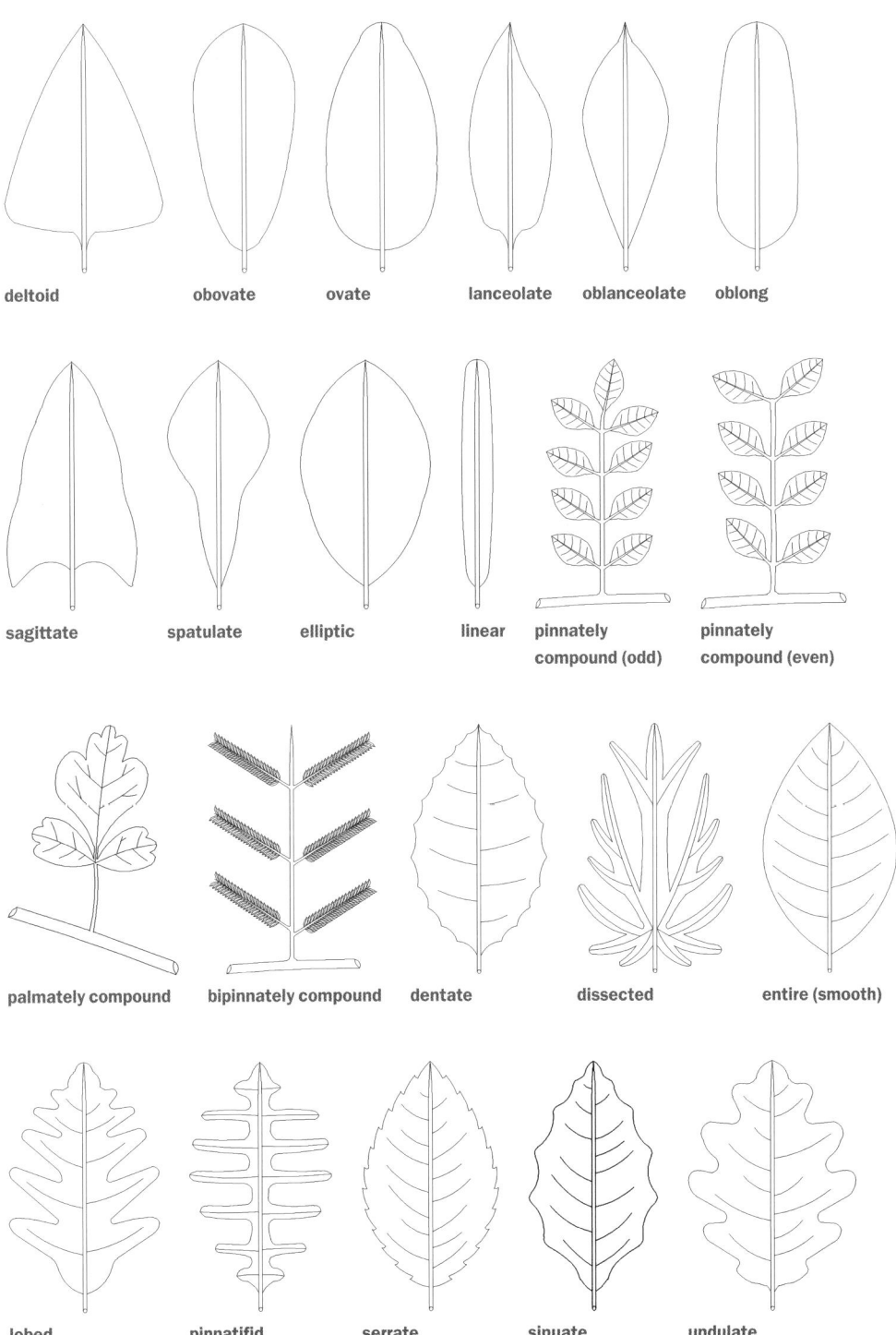

# Appendix B. Tank Mixing Guide

## Chemical Mixing Calculations—Individual Plant Treatments

| Tank size Gallons | ¼% Ounces | ½% Ounces | ¾% Ounces | 1% Ounces | 1½% Ounces | 4% Ounces | How many tanks will a 2.5-gallon container make at 1%? No. of tanks |
|---|---|---|---|---|---|---|---|
| 1   | ⅓    | ⅔    | 1     | 1⅓    | 2     | 5     | 250.00 |
| 3   | 1    | 2    | 3     | 4     | 6     | 15½   | 83.33  |
| 4   | 1⅓   | 2½   | 4     | 5     | 7½    | 20½   | 62.50  |
| 5   | 1⅔   | 3⅓   | 5     | 6½    | 10    | 26    | 50.00  |
| 10  | 3⅓   | 6½   | 10    | 13    | 19    | 51    | 25.00  |
| 14  | 4⅓   | 9    | 13⅓   | 18    | 27    | 71½   | 17.86  |
| 15  | 4⅔   | 9½   | 14⅓   | 19¼   | 29    | 77    | 16.67  |
| 25  | 8    | 16   | 24    | 32    | 48    | 128   | 10.00  |
| 35  | 11⅓  | 22½  | 33⅔   | 44¾   | 67    | 179   | 7.14   |
| 50  | 16   | 32   | 48    | 64    | 96    | 256   | 5.00   |
| 55  | 17⅔  | 35¼  | 52⅔   | 70½   | 105½  | 281½  | 4.55   |
| 65  | 20⅔  | 41½  | 62⅓   | 83¼   | 125   | 333   | 3.85   |
| 85  | 27⅓  | 54½  | 81⅔   | 108¾  | 163   | 435   | 2.94   |
| 100 | 32   | 64   | 96    | 128   | 192   | 512   | 2.50   |
| 110 | 35⅓  | 70½  | 105⅔  | 140¾  | 211   | 563   | 2.27   |
| 125 | 40   | 80   | 120   | 160   | 240   | 640   | 2.00   |
| 225 | 72   | 144  | 216   | 288   | 432   | 1152  | 1.11   |

How to use this chart: If you have a 50-gallon tank and need a 1% concentration of chemical solution, then you would add 64 ounces of chemical to your 50-gallon tank. If your product also required a ¼% concentration of surfactant and a ¼% concentration of dye, then you would also add 16 ounces of surfactant and 16 ounces of dye to your 50-gallon tank.

Always follow the directions on the product label for correct chemical tank mix recommendations.

# Appendix C. Common Conversions

## Area
1 acre = 43,560 square feet = 209 feet x 209 feet = 69.5 yards x 69.5 yards
1 section = 640 acres = 1 square mile
1 hectare = 2.471 acres
1 square foot = 144 square inches
1 square yard = 9 square feet

## Length
1 mile = 5,280 feet = 1,760 yards = 1.61 kilometers
1 rod = 16.5 feet
1 chain = 66 feet
1 kilometer = 0.62 miles
1 roll of barbed wire = ¼ mile = 1,320 feet
1 roll of net wire = 330 feet = 20 rods

## Weight
1 pound = 453.6 grams = 0.45 kilograms = 16 ounces
1 gram = 0.035 ounces
1 ounce = 28.35 g
1 short ton = 2,000 pounds
1 long ton = 2,240 pounds
1 kilogram = 2.2 pounds

## Liquid Measure
1 gallon (gal) = 4 quarts (qt) = 8 pints (pt) = 16 cups (c) = 128 fluid ounces (fl oz)
1 quart = 2 pints = 4 cups = 32 fluid ounces
1 pint = 2 cups = 16 fluid ounces
1 cup = 8 fluid ounces
1 fluid ounce = 2 tablespoons (tbs)
1 tablespoon (tbs) = 3 teaspoons (tsp) = ½ fluid ounce
1 teaspoon = ⅙ fluid ounce

## Calculation of Water Storage Capacity:
Round tank (gallons) = 3.1416 x radius squared (ft.) x height (ft.) x 7.48
Rectangular tank (gallons) = height (ft.) x width (ft.) x length (ft.) x 7.48
1 gallon = 8.355 pounds
1 cubic foot of water = 62.43 pounds or 7.48 gallons
1 acre inch of water = 27,154 gallons

# Appendix D. Livestock Husbandry

**Approximate Peak Water Requirements**
Cattle = 7 to 16 gallons/day (West Texas use 15 gallons/day)
Horses = 8 to 12 gallons/day
Sheep and goats = 1 to 4 gallons/day

**Approximate Gestation Periods**
Cattle = 283 days
Horses = 336 days
Sheep = 148 days
Goats = 151 days

**Approximate Maximum Distance Livestock Travel to Water**
Terrain Miles:
Rough: 0.5
Rolling, hilly: 1.0
Smooth, flat: 2.0
Sandy: 1.5
Undulating dunes: 1.0

# Glossary

**Air-dry weight**—The weight of a substance, usually vegetation, after it has been allowed to dry to equilibrium with the atmosphere. This is typically the base unit used when conducting rangeland inventories of vegetation in order to determine available forage and stocking rates.

**Alternate**—In reference to plant leaves, growing on one side of the stem and then the other, but not directly across from each other.

**Animal Unit (AU)**—One mature (1,000 lb.) animal or the equivalent based on an average air-dry forage consumption of 30 lbs. per day.

**Animal Unit Day (AUD)**—Amount of air-dry forage required by an animal unit for one day (30 lbs).

**Animal Unit Equivalent (AUE)**—A numerical figure expressing the forage requirements of a particular kind and class of animal relative to the requirements for one animal unit. See the AUE chart under "Overview of Livestock Grazing Management" in chapter 5.

**Animal Unit Month (AUM)**—Amount of air-dry forage required by an animal unit for one month (900 lbs).

**Animal Unit Year (AUY)**—Amount of air-dry forage required by an animal unit for one year (10,950 lbs).

**Annual plant**—A plant that completes its life cycle in one year and in which new plants sprout from seeds each year.

**Available forage**—That portion of the forage production that is accessible for use by a specified kind or class of grazing animal.

**Awn**—In reference to a plant, a hairlike growth, especially on parts of a seed.

**Basal**—Referring to the base or lower portion of a structure.

**Browse**—The leaves and tender twigs of woody plants that are used for forage.

**Browse line**—The physical structure and appearance of taller shrubs and trees subjected to heavy browsing by the removal of branch, twig, and leaf growth within the reach of an animal. See photo example under "Reading the Landscape" in chapter 5.

**Brush**—Woody plants that increase to undesirable levels and are not considered important forage plants.

**Canopy**—The aboveground, aerial portion of vegetation, usually associated with shrubs and trees. Usually expressed in percentage of ground covered (canopy cover).

**Carrying capacity**—Maximum stocking rate that is possible year to year without inducing damage to vegetation or related resources; often confused with stocking rate.

**Community**—A particular area or portion of land that has a particular combination of vegetation. Some plant species like to grow alongside other plant species in the same area or community.

**Cool-season plant**—A plant that completes most of its growth in late fall, winter, or early spring.

**Cropland**—Land used primarily for the production of cultivated crops.

**Culm**—The stem or stalk of a grass.

**Ecological site**—A distinctive kind of land with specific physical characteristics that differs from other kinds of land in its ability to produce a distinctive kind and amount of vegetation.

**Ecology**—The study of the interrelationships of organisms with their environment.

**Foliage**—A collective term for the green or live leaves of plants.

**Forage**—All herbaceous vegetation and browse that is used as food by grazing animals. The term "to forage" is similar to the term "to graze."

**Forb**—A mostly herbaceous plant, excluding grasses and grasslike plants.

**Grazing**—The consumption of forage by livestock or wildlife.

**Hedging**—The physical structure and appearance of a woody plant that develops over a period of years when terminal twigs are browsed, causing lateral twigs to develop. Moderate hedging is not harmful to a plant, since more leaf surface is retained and more of the new production of browse is still within reach of browsing animals. Severe hedging causes more stubby branches with limited leaf production, thereby reducing plant health and greatly reducing browse production.

**Historical climax plant community**—A plant community that was naturally best adapted to a particular ecological site in North America at the time of European immigration and settlement. A range professional often compares the current plant communities of a pasture with the historical communities for that site.

**Infiltration**—The intake of water into the soil profile. Healthy rangeland with abundant native grasses has better water infiltration and thus better soil moisture than degraded rangeland.

**Inflorescence**—The seedhead or flowering part of a grass plant.

**Introduced plant**—A plant that is brought in from outside North America. Texas may have plants that are native to North America yet are not historically from Texas.

**Key species**—One or two plant species chosen to serve as a guide to the grazing use of an entire plant community. If the key species on a key grazing area is properly grazed, then the entire plant community will not be excessively grazed.

**Motte**—A group of trees or shrubs in an otherwise open landscape area. See "Reading the Landscape" in chapter 5 for a full description.

**Native plant**—In a broad sense, a plant is native if it is found in North America and grew there historically. However, depending on the usage of the term, a plant may be native to North America but not native to Texas. Also, a plant may be native to Texas but not to a region like the Concho Valley. See also "Introduced plant."

**Node**—A joint or knot on the stem of a grass, sometimes enlarged or darkened; often used to help identify a plant.

**Pastureland**—Land where grasses are managed for grazing. These lands receive periodic treatments such as tillage, fertilization, mowing, and weed control and may be irrigated or dry. These lands are not in crop rotation.

**Pedicel**—Tiny stem of an individual flower or spikelet.

**Perennial plant**—A plant that lives for three or more years, growing back each year from root stalks, crown buds, or branches.

**Photosensitization**—Increased sensitivity to sunlight, possibly resulting in skin inflammation and sunburns; often caused by consumption of particular plants.

**Pistil**—The female reproductive organ of a flower, consisting of an ovary, stigma, and style.

**Prostrate**—Describes a plant lying flat or growing on the ground.

**Pubescent**—In reference to a plant, having hairs.

**Rangeland**—Land on which the historical climax plant community was predominantly grasses, grasslike plants, forbs, or shrubs. Rangelands include natural grasslands, savannas, shrublands, most deserts, tundra, alpine communities, coastal marshes, and wet meadows. Locally, "rangeland" describes uncultivated land used for grazing and wildlife.

**Rangeland health**—The degree to which the integrity of the soil, vegetation, water, and air as well as the ecological processes of a rangeland ecosystem are balanced and sustained.

**Rangeland trend**—The direction of change in an existing plant community relative to the historical climax plant community for an ecological site.

**Range seeding**—The process of establishing vegetation by applying seeds to rangeland.

**Range site**—See "Ecological site."

**Rhizome**—Underground stems that extend horizontally and may form roots and new plants at the nodes.

**Riparian**—The area immediately adjacent to streams, lakes, or other natural water bodies. A transitional area between water bodies and uplands.

**Scarification**—The process of mechanically or chemically treating a seed coat in order to improve moisture absorption and thereby enhance germination.

**Spikelet**—The basic unit of a grass seedhead or inflorescence.

**Stigma**—Portion of the pistil that is used to catch or receive pollen.

**Stocking density**—The relationship between number of animals and area of land at any instant in time.

**Stocking rate**—A ratio of the number of specific kinds and classes of animals grazing or utilizing a unit of land for a specific period of time. May be expressed as animals per acre (or per section) or acres per animal.

**Stolon**—Aboveground horizontal runners that may take root at the nodes. See also "Rhizome."

**Style**—Middle portion of a pistil, between the ovary and stigma, often long and narrow.

**Tufted**—Describes a plant, usually a grass, growing in clumps from the base, usually without runners.

**Warm-season plant**—A plant that completes most of its growth in summer and fall.

**Weed**—Any plant that grows in an area where it is unwanted. Forbs are often mistakenly called weeds. By this definition, even a grass can be a weed if it grows where it is not wanted.

# References

Ajilvsgi, G. 2003. *Wildflowers of Texas.* Rev. ed. Fredericksburg, TX: Shearer Publishing.

Amos, B. 1998. *Checklist of the Plants of the Concho Valley of Texas.* San Angelo, TX: Angelo State University Department of Biology.

Anderson, R. C. 1990. "The Historic Role of Fire in the North American Grassland." In *Fire in North American Tallgrass Prairies,* edited by S. L. Collins and L. L. Wallace, 8–18. Norman: University of Oklahoma Press.

Archer, S. 1994. "Woody Plant Encroachment into Southwestern Grasslands and Savannas: Rates, Patterns, and Proximate Causes." In *Ecological Implications of Livestock Herbivory in the West,* edited by M. Vavra, W. A. Laycock, and R. D. Pieper, 13–68. Denver: Society for Range Management.

Cadenhead, J. F., A. McGinty, C. W. Hanselka, B. Lyons, and C. R. Hart. 2005. *How to Control Common (Annual) Broomweed.* L- 5461. College Station: Texas AgriLife Extension Service, Texas A&M University.

Carlson, P. H. 1982. *Texas Woollybacks: The Range Sheep and Goat Industry.* College Station: Texas A&M University Press.

Copeland, R. T. 1978. "Grasslands." *Encyclopedia Britannica (Macropedia).* 14th ed. 8:280–86. Chicago: William Benton.

Cory, V. L. 1949. "On Some Grasses, Chiefly of the Edwards Plateau of Texas." *Field and Laboratory* 17:41–52.

Diggs, G. M., B. L. Lipscomb, and R. J. O'Kennon. 1999. *Shinners and Mahler's Illustrated Flora of North Central Texas.* Fort Worth: Botanical Research Institute of Texas.

Enquist, M. 1987. *Wildflowers of the Texas Hill Country.* Austin, TX: Lone Star Botanical.

Everitt, J. H., and D. L. Drawe. 1993. *Trees, Shrubs and Cacti of South Texas.* Lubbock: Texas Tech University Press.

Foster, J.H. 1917. *The Spread of Timbered Areas in Central Texas.* Journal of Forestry 15:442-445.

Froebel, J. 1859. *Seven Years' Travel in Central America, Northern Mexico, and the Far West of the United States.* London: Richard Bentley.

Gilmore, M. R. 1919. *Use of Plants by the Indians of the Missouri River Region.* Thirty-Third Annual Report of the Bureau of American Ethnology to the Secretary of the Smithsonian Institution, 1911–1912. Washington, DC: Government Printing Office.

Gleason, H. 1923. "The Vegetational History of the Middle West." *Annals of the Association of American Geographers* 12:39–85.

Gould, F. 1975. *The Grasses of Texas.* College Station: Texas A&M University Press.

———. 1978. *Common Texas Grasses: An Illustrated Guide.* College Station: Texas A&M University Press.

Gray, A. B. 1856. *Survey of a Route for the Southern Pacific R. R., on the 32nd Parallel.* Cincinnati: Wrightson.

Haines, F. 1970. *The Buffalo.* Norman: University of Oklahoma Press.

Haley, J. F. 1929. "Grass Fires of the Southern Plains." In *West Texas Historical Association Year Book*, 24–42.

Hart, C. R., T. Garland, C. Barr, B. B. Carpenter, and J. C. Reagor. 2010. *Toxic Plants of Texas.* B-6105. College Station: Texas AgriLife Extension Service, Texas A&M University.

Hatch, S. L., and J. P. Pluhar. 1993. *Texas Range Plants.* College Station: Texas A&M University Press.

Hunt, W. R., and J. Leffler. 2010. "Irion County." *Handbook of Texas Online.* Texas State Historical Association. Accessed June 25, 2015. http://www.tshaonline.org/handbook/online/articles/hci01.

Jenkins, R. B. 1939. "Cedars and Poverty." *The Cattleman* (June): 61–62.

Jones, S. D., J. K. Wipff, and P. M. Montgomery. 1997. *Vascular Plants of Texas.* Austin: University of Texas Press.

Kirkpatrick, Z. M. 1992. *Wildflowers of the Western Plains.* Austin: University of Texas Press.

Kneuper, C. L., C. B. Scott, and W. E. Pinchak. 2003. "Consumption and Dispersion of Mesquite Seeds by Ruminants." *Journal of Range Management* 56:255–59.

Lower Colorado River Authority. 2008. *Colorado River Basin Highlights Report.*

Martin, P. S. 1975. "Vanishings, and Future, of the Prairie." *Geoscience and Man* 10:39–49.

Maxwell, T. C. 1979. "Avifauna of the Concho Valley of West-Central Texas with Special Reference to Historical Change." PhD diss., Texas A&M University.

McGinty, A., J. Ansley, W. Hamilton, C. Hart, L. Redmon, and R. Lyons. 2010. *Chemical Weed and Brush Control Suggestions for Rangeland.* B-1466. College Station: Texas AgriLife Extension Service, Texas A&M University.

Moore, M. 1989. *Medicinal Plants of the Desert and Canyon West.* Santa Fe: Museum of New Mexico Press.

———. 2003. *Medicinal Plants of the Mountain West.* Santa Fe: Museum of New Mexico Press.

Nokes, J. 2001. *How to Grow Native Plants of Texas and the Southwest.* Austin: University of Texas Press.

Olmsted, F. L. 1857. *A Journey through Texas or, a Saddle-Trip on the Southwestern Frontier.* Austin: University of Texas Press.

Ormsby, W. L. 1942. *The Butterfield Overland Mail: Only Through Passenger on the First Westbound Stage.* San Marino, CA: Henry E. Huntington Library and Art Gallery.

Parker, A. A. 1836. *Trip to the West and Texas, Comprising a Journey of 8,000 Miles, through New York, Michigan, Illinois, Missouri, Louisiana and Texas in the Autumn and Winter of 1834–35.* 2nd ed. Concord, NH: William White.

Powell, A. M. 1994. *Grasses of the Trans-Pecos and Adjacent Areas.* Austin: University of Texas Press.

Rector, B. 2003. *Know Your Grasses.* College Station: Texas Cooperative Extension, Texas A&M University.

Sanders, C. F. 1920. *Useful Wild Plants of the United States and Canada.* New York: Robert M. McBride.

Smeins, F. E. 1980. "Natural Role of Fire on the Edwards Plateau of Texas." In *Prescribed Burning of the Edwards Plateau of Texas,* edited by L. D. White, 4–16. College Station: Texas Agricultural Extension Service.

———. 1984. "Origin of the Brush Problem: A Geological and Ecological Perspective of Contemporary Distributions." In *Proceedings of Brush Management Symposium,* edited by K. W. McDaniel, 5–16. Lubbock: Texas Tech Press.

Smith, J. G. 1899. *Grazing Problems in the Southwest and How to Meet Them.* USDA Bulletin No. 16. Washington, DC: Government Printing Office.

Stevenson, M. C. 1908. *Ethnobotany of the Zuni Indians.* Thirtieth Annual Report of the Bureau of American Ethnology to the Secretary of the Smithsonian Institution, 1908–1909. Washington, DC: Government Printing Office.

Tellman, B. 1998. "Stowaways and Invited Guests: How Some Exotic Plant Species Reached the American Southwest." In *The Future of Arid Grasslands: Identifying Issues, Seeking Solutions,* edited by B. Tellman, D. Finch, C. Edminster, and R. Hamre, 144–49. Proceedings RMRS-P-3. Fort Collins, CO: USDA Forest Service, Rocky Mountain Research Station.

Thurston, E. L. 1976. "Morphology, Fine Structure and Ontogeny of the Stinging Emergence of *Tragia ramosa* and *T. saxicola* (Euphorbiaceae). *American Journal of Botany* 63 (July): 710–18.

Tull, D. 1987. *Edible and Useful Plants of Texas and the Southwest.* Austin: University of Texas Press.

USDA (United States Department of Agriculture) Natural Resources Conservation Service. 1994. *The Use and Management of Browse in the Edwards Plateau of Texas.* Temple, TX.

US EPA (United States Environmental Protection Agency). 2005. "Citronellol (167004) Fact Sheet." http://www.epa.gov/opp00001/chem_search/reg_actions/registration/fs_PC-167004_22-Apr-04.pdf.

Vines, R. A. 1960. *Trees, Shrubs, and Woody Vines of the Southwest.* Austin: University of Texas Press.

Webb, W. P. 1931. *The Great Plains.* New York: Grossett and Dunlap.

———. 1935. *The Texas Rangers: A Century of Frontier Defense.* Boston: Houghton Mifflin.

Wilcox, B. P. 2002. "Shrub Control and Streamflow on Rangelands: A Process-Based Viewpoint." *Journal of Range Management* 55:318–26.

Wilcox, B. P., and Y. Huang. 2010. "Woody Plant Encroachment Paradox: Rivers Rebound as Degraded Grasslands Convert to Woodlands." *Geophysical Research Letters* 37. L07402, doi:10.1029/2009/ GL041929.

South Concho River in Christoval, Texas. Photo by George Clendenin, USDA-NRCS.

# Index of Plants (Common and Scientific Names)

Note: Page numbers in italic indicate figures or photos. Page numbers in bold indicate maps.

*Abutilon fruticosum*, 26, *270, 271*, 557
*Acacia greggii*, 27, *366*, 367, **479**
*Acacia roemeriana*, 27, *368*, 369, **479**
*Acalypha phleoides*, 25, *204, 205*
Acanthaceae: *Ruellia metziae*, 24, *116*, 117, 557; *Siphonoglossa pilosella*, 24, *118*, 119
*Acleisanthes longiflora*, 26, *282, 283*
aflatoxins, 546
Agavaceae: *Nolina texana*, 29, *456*, 457; *Yucca constricta*, 29, *458*, 459–61, *460*
agaves and cacti, 29. See also *specific plants by scientific or common name*
Agriculture, U.S. Department of (USDA), 10, 15
algerita, 27, *348*, 349, **500**, 552, 556
*Allium drummondii*, 23, *36*, 37, **503**, 557
*Aloysia gratissima*, 28, *426, 427*, **502**
*Amaranthus* spp., 556
*Amblyolepis setigera*, 24, *128*, 129, **502**
*Ambrosia confertiflora*, 24, *130*, 131, **477**, **503**
*Ambrosia psilostachya*, 24, *132*, 133, **477**, **503**
American elm, 28, *424, 425*, **508**
American Indians, use of fire by, 516, 518
Anacardiaceae: *Rhus copallinum*, 27, *338*, 339, **480**, **500**; *Rhus microphylla*, 27, *340*, 341, **500**, 552; *Rhus trilobata*, 27, *342*, 343, **500**, **502**; *Rhus virens*, 27, *344*, 345; *Toxicodendron radicans*, 28, *430*, 431, 556
*Andropogon gerardii*, 23, *40*, 41
*Andropogon glomeratus*, 23, *42*, 43, **504**
angel trumpets, 26, *282, 283*
Anglo Texan settlers, 518–22

animals: browse lines created by, 497; carrying capacity for, 494; foraging behavior of, 13–14, 495, 496; toxic plant poisoning of, 13–14. See also wildlife management; *specific animals by name*
animal unit (AU), 495
animal unit equivalent (AUE), 495, 496
antelope horn milkweed, 24, *124, 125*
*Aphanostephus skirrhobasis*, 24, *134*, 135
Apiaceae, 24. See also *Eryngium leavenworthii*
Arden, John, 493
*Argemone aurantiaca*, 26, *300*, 301, 557
*Aristida purpurea* var. *longiseta*, 23, *44*, 45
*Aristida purpurea* var. *wrightii*, 23, *46*, 47
Aristolochiaceae, 24. See also *Aristolochia coryi*
*Aristolochia coryi*, 24, *122*, 123
*Artemisia ludoviciana*, 24, *136*, 137, **502**
Asclepiadaceae: *Asclepias asperula*, 24, *124, 125*; *Asclepias latifolia*, 24, *126, 127*; *Funastrum crispum*, 28, *432*, 433
*Asclepias asperula*, 24, *124*, 125
*Asclepias latifolia*, 24, *126*, 127
Ashe juniper: classification of, 27; management of, 495, **500**, 527; profile of, *354*, 355–57, *356*; smell emitted by, **502**
Asteraceae: *Amblyolepis setigera*, 24, *128*, 129, **502**; *Ambrosia confertiflora*, 24, *130*, 131, **477**, **503**; *Ambrosia psilostachya*, 24, *132*, 133, **477**, **503**; *Aphanostephus skirrhobasis*, 24, *134, 135*; *Artemisia ludoviciana*, 24, *136*, 137,

Asteraceae (*cont.*)
502; *Aster ericoides*, 24, *138*, 139; *Baccharis neglecta*, 27, *346*, 347; *Engelmannia peristenia*, 24, *140*, 141, 557; *Erigeron strigosus*, 24, *142*, 143; *Evax verna*, 24, *144*, 145; *Gaillardia pinnatifida*, 24, *146*, 147; *Gaillardia pulchella*, 24, *148*, 149; *Grindelia nuda*, 24, *150*, 151; *Grindelia papposa*, 24, *152*, 153; *Gutierrezia dracunculoides*, 24, *154*, 155, 557; *Gutierrezia sarothrae*, 14, 24, *156*, 157–59, *158*, 502; *Gymnosperma glutinosum*, 24, *160*, 161; *Heterotheca canescens*, 24, *162*, 163; *Hymenopappus artemisiifolius*, 24, *164*, 165; *Liatris punctata*, 24, *166*, 167, *510*, 557; *Lygodesmia texana*, 24, *168*, 169; *Machaeranthera pinnatifida*, 24, *170*, 171; *Melampodium leucanthum*, 24, *172*, 173; *Parthenium confertum*, 24, *174*, 175; *Ratibida columnifera*, 24, *176*, 177, *564*; *Senecio douglasii*, 24, *178*, 179; *Simsia calva*, 24, *180*, 181, *478*; *Solidago canadensis*, 25, *182*, 183, 557; *Thelesperma filifolium*, 25, *184*, 185; *Thymophylla pentachaeta*, 25, *186*, 187; *Verbesina encelioides*, 25, *188*, 189, 557; *Wedelia texana*, 25, *190*, 191, 478; *Zinnia grandiflora*, 25, *192*, 193
*Aster ericoides*, 24, *138*, 139
*Astragalus nuttallianus*, 25, 226, 227
*Atriplex canescens*, 27, *352*, 353
AU (animal unit), 495
AUE (animal unit equivalent), 495, *496*
Austin, Stephen F., 517
axis deer, 506, 540

*Baccharis neglecta*, 27, *346*, 347
back fires, 525
baiting laws, 548
balsam gourd, 28, *444*, 445
barnyard grass, 557
Bartlett, John R., 487, 490–91

beak shape of birds, 556
bean family. *See* Fabaceae
bellflower family. *See* Campanulaceae
Berberidaceae, 27. See also *Berberis trifoliolata*
*Berberis trifoliolata*, 27, *348*, 349, 500, 552, 556
big bluestem, 23, *40*, 41
birds, 554–61. *See also* songbirds; specific birds by name
bird-watching, 554
birthwort family. *See* Aristolochiaceae
bison. *See* buffalo
black willow, 28, *412*, 413, 504
bladderpod sida, 26, *274*, 275, 557
blazingstar family. *See* Loasaceae
bluebell family. *See* Campanulaceae
blue curls, 7
blue quail, 550
bluets, 26, *310*, 311
bobwhite quail, 550
*Boerhavia linearifolia*, 26, *284*, 285
Boraginaceae, 25. See also *Tiquilia canescens*
*Bothriochloa laguroides*, 23, *48*, 49
bottomlands. *See* riparian areas
*Bouteloua curtipendula*, 23, *50*, 51, 551
*Bouteloua hirsuta*, 23, *52*, 53
*Bouteloua rigidiseta*, 23, *54*, 55
*Bouteloua trifida*, 23, *56*, 57
Brady's Creek, 488, 490
Brassicaceae, 25. See also *Lepidium virginicum*
broadleaf milkweed, 24, *126*, 127
*Bromus catharticus*, 23, *58*, 59
broom snakeweed, 14, 24, *156*, 157–59, *158*, 502
brown-headed cowbird, 561
browse, as nutritional source, 539, 541
browse lines, 497–98
brush management, 540–41
Bryan, F. T., 487, 488–90
*Buchloe dactyloides*, 23, *60*, 61
Buckley yucca, 29, *458*, 459–61, *460*
buckthorn family. *See* Rhamnaceae

buffalo, 488, 490, 493, 516, 517
buffalobur, 26, *324*, 325
buffalo gourd, 28, *442*, 443, 502
buffalograss, 23, *60*, 61
bumella, 500
bunchgrasses, 503–4
bur clover, 25, *236*, 237
burning, prescribed, *501*, 503, 518, 524–29, *528–34*
bur oak, 27, *380*, 381, *508*
bush sunflower, 24, *180*, 181, 478
bushy bluestem, 23, *42*, 43, 504
buttercup family. *See* Ranunculaceae
butterflies, *562–64*
buttonbush, 28, *408*, 409, 504

Cactaceae: *Echinocactus texensis*, 29, *462*, 463; *Echinocereus enneacanthus*, 29, *464*, 465; *Echinocereus reichenbachii*, 29, *466*, 467; *Opuntia engelmannii* (*See* prickly pear); *Opuntia leptocaulis*, 29, *472*, 473, 528, 556
cacti and agaves, 29. See also *specific plants by scientific or common name*
calcium carbonate, 10
California filaree, 25, *248*, 249
*Callirhoe involucrata*, 26, *272*, 273, 557
caltrop family. *See* Zygophyllaceae
*Calylophus hartwegii*, 26, *290*, 291
Campanulaceae, 25. See also *Lobelia cardinalis*
Canada wildrye, 23, *68*, 69
Caprifoliaceae, 27. See also *Lonicera albiflora*
cardinal flower, 25, *198*, 199, *507*
cardinals, *555*
*Carex planostachys*, 23, *32*, 33
Carolina snailseed, 28, *446*, 447, 556
carrot family. *See* Apiaceae
carrying capacity, 494
*Carya illinoinensis*. *See* pecan
catclaw acacia, 27, *366*, 367, 479
catclaw mimosa, 27, *372*, 373, 479
cattle, 540. See also longhorn cattle
cedar sedge, 23, *32*, 33

*Celtis laevigata* var. *reticulata*, 28, *422*, 423
*Centaurea melitensis*, 16
*Centaurea solstitialis*, 16
*Centaurium beyrichii*, 25, *246*, 247
Central Edwards Plateau, 8, **9**
century-plant family. *See* Agavaceae
*Cephalanthus occidentalis*, 28, *408*, 409, 504
Chadbourne, Fort, 491
Chalk Hill woolly-white, 24, *164*, 165
*Chamaesyce lata*, 25, *206*, 207
Chenopodiaceae: *Atriplex canescens*, 27, *352*, 353; *Salsola tragus*, 25, *200*, 201–3, *202*
*Chenopodium album*, 557
chicle family. *See* Sapotaceae
Childress, Billy, 493
*Chloris cucullata*, 23, *62*, 63
*Chloris verticillata*, 23, *64*, 65
*Cissus incisa*, 28, *452*, 453, 556
citrus family. *See* Rutaceae
citrus-smelling plants, 502
*Clematis drummondii*, 28, *448*, 449
Clendenin, George, 18
climate, of Concho Valley, 4, 6–7, *8*
climax vegetation theory, 485
*Cocculus carolinus*, 28, *446*, 447, 556
coffee family. *See* Rubiaceae
Colorado River Basin, 4, **5**
common broomweed, 24, *154*, 155, 557
common dogweed, 25, *186*, 187
common horehound, 25, *256*, 257, 502
Concho River watershed, 4, **5**
Concho Valley, 3–11; climatic data in, 4, 6–7, *8*; geographic characteristics of, 3, 4, **5;** historical accounts of, 485–93; Major Land Resource Areas of, 8–10, **9**; precipitation in, 4, 6–7, *8*; reading landscape in, 497–504; riparian areas in, 7, *484*, 504–6, *507–12*; rivers and streams of, **486**, *507*; soils in, 10, **11**, *12*, 504; vegetation of, 485–93. *See also* landscape features
*Condalia ericoides*, 27, *402*, 403

*Condalia hookeri,* 27, *404,* 405, *500,* 556
Convolvulaceae: *Convolvulus arvensis,* 28, *434,* 435; *Convolvulus equitans,* 28, *436,* 437; *Ipomoea cordatotriloba,* 28, *438,* 439; *Ipomoea lindheimeri,* 28, *440,* 441
*Convolvulus arvensis,* 28, *434,* 435
*Convolvulus equitans,* 28, *436,* 437
*Cooperia drummondii,* 23, *38,* 39
Cortés, Hernando, 517
Cory, V. L., 520
cotton morning glory, 28, *438,* 439
cowpen daisy, 25, *188,* 189, 557
*Croton dioicus,* 25, *208,* 209, 503
*Croton monanthogynus,* 25, *210,* 211, 503
*Croton pottsii,* 25, *212,* 213, 503
Cucurbitaceae: *Cucurbita foetidissima,* 28, *442,* 443, 502; *Ibervillea lindheimeri,* 28, *444,* 445
*Cucurbita foetidissima,* 28, *442,* 443, 502
culling practices, 538
culture, fire, 523–26, *534*
Cupressaceae. See Ashe juniper; redberry juniper
curlycup gumweed, 24, *150,* 151
curly-mesquite, 23, *78,* 79
cutleaf daisy, 24, *170,* 171
Cyperaceae: *Carex planostachys,* 23, *32,* 33; *Eleocharis* spp., 23, *34,* 35, 504
cypress family. See Ashe juniper; redberry juniper

daisy family. See Asteraceae
*Dalea formosa,* 27, *370,* 371
*Dalea lasiathera,* 25, *228,* 229
*Dalea nana,* 25, *230,* 231
*Datura stramonium,* 3, 26, *316,* 317
deer. See axis deer; white-tailed deer
deer pea vetch, 25, *244,* 245
*Desmanthus velutinus,* 25, *232,* 233
de Soto, Hernando, 517
destocking, 495
dewberry, 556
diet and nutrition: doves, 548–49; quail, 552, *553;* songbirds, 556–57, *562;* turkeys, 546, *547;* white-tailed deer, 538–40, 541–42, *542–44*
*Digitaria cognata,* 23, *66,* 67
*Diospyros texana,* 27, *362,* 363, 500
*Ditaxis humilis,* 25, *214,* 215, 556–57
double fireguards, *533*
Dove Creek, 4, **5,** 489, 490, 506–7
doves, 548–50, *549*
drought, destocking during, 495
Drummond's oxalis, 26, *296,* 297
Dutchman's breeches, 26, *312,* 313, 502
Dutchman's pipe, 24, *122,* 123
dwarf dalea, 25, *230,* 231

eastern gamagrass, 24, *114,* 115, 504, 510
Ebenaceae, 27. See also *Diospyros texana*
*Echinocactus texensis,* 29, *462,* 463
*Echinocereus enneacanthus,* 29, *464,* 465
*Echinocereus reichenbachii,* 29, *466,* 467
*Echinochloa crus-galli,* 557
Ecological Site Descriptions (ESDs), 10
ecosystems, defined, 16
ecotourism, 554
Ector soils, 10, **11,** 12
Edwards Plateau: Central, 8, **9;** fireproofing of, 523, *524;* prescribed burning on, 526; ranchers of, 493, 522; vegetation of, 516–18; Western, 8, **9,** 515, 519, 522
elbowbush, 27, *400,* 401, 556
*Eleocharis* spp., 23, *34,* 35, 504
elm family. See Ulmaceae
*Elymus canadensis,* 23, *68,* 69
*Elymus longifolius,* 23, *70,* 71
Emory sedge, 504
Engelmann daisy, 24, *140,* 141, 557
*Engelmannia peristenia,* 24, *140,* 141, 557
ephedra, 27, *364,* 365, 499
*Ephedra antisyphilitica,* 27, *364,* 365, 499

Ephedraceae, 27. See also *Ephedra antisyphilitica*
*Eragrostis intermedia*, 23, 72, 73
*Erigeron strigosus*, 24, 142, 143
*Eriochloa sericea*, 23, 74, 75
*Erioneuron pilosum*, 23, 76, 77
*Erodium cicutarium*, 25, 248, 249
*Erodium texanum*, 25, 250, 251
*Eryngium leavenworthii*, 24, 120, 121, 557
ESDs (Ecological Site Descriptions), 10
ethnobotany, 17
Euphorbiaceae: *Acalypha phleoides*, 25, 204, 205; *Chamaesyce lata*, 25, 206, 207; *Croton dioicus*, 25, 208, 209, 503; *Croton monanthogynus*, 25, 210, 211, 503; *Croton pottsii*, 25, 212, 213, 503; *Ditaxis humilis*, 25, 214, 215, 556–57; *Euphorbia marginata*, 25, 216, 217; *Phyllanthus polygonoides*, 25, 218, 219, 557; *Stillingia texana*, 25, 220, 221; *Stillingia treculiana*, 25, 222, 223; *Tragia ramosa*, 25, 224, 225, 557
euphorbia family. See Euphorbiaceae
*Euphorbia marginata*, 25, 216, 217
Eurasian collared doves, 549–50
*Evax verna*, 24, 144, 145
evening primrose family. See Onagraceae
evergreen sumac, 27, 344, 345

Fabaceae: *Acacia greggii*, 27, 366, 367, 479; *Acacia roemeriana*, 27, 368, 369, 479; *Astragalus nuttallianus*, 25, 226, 227; *Dalea formosa*, 27, 370, 371; *Dalea lasiathera*, 25, 228, 229; *Dalea nana*, 25, 230, 231; *Desmanthus velutinus*, 25, 232, 233; *Lupinus subcarnosus*, 25, 234, 235, 557; *Medicago minima*, 25, 236, 237; *Mimosa aculeaticarpa* var. *biuncifera*, 27, 372, 373, 479; *Mimosa borealis*, 27, 374, 375, 479; *Mimosa nuttallii*, 25, 238, 239; *Prosopis glandulosa*, 27, 376, 377–79, 378, 493, 519, 556; *Rhynchosia senna*, 25, 240, 241; *Senna roemeriana*, 25, 242, 243; *Vicia ludoviciana*, 25, 244, 245
Fagaceae: *Quercus macrocarpa*, 27, 380, 381, 508; *Quercus mohriana*, 27, 382, 383–85, 384; *Quercus pungens* var. *vaseyana*, 27, 386, 387, 389; *Quercus sinuata* var. *breviloba*, 14, 27, 390, 391; *Quercus virginiana* (*See* live oak)
fall witchgrass, 23, 66, 67
false pennyroyal, 25, 254, 255, 502
fawning cover, 541
feather dalea, 27, 370, 371
Federal List of Noxious Plants, 14
fences, wolf-proof, 523
feral hogs, 506, 521
field bindweed, 28, 434, 435
field ragweed, 24, 130, 131, 477, 503
figwort family. See Scrophulariaceae
fire culture, 523–26, 534
fireguards, 517–18, 528–29, 531, 533
fires, 513–26; Anglo Texan settlers and, 518–22; current use of, 526; Edward Plateau vegetation and, 516–18; fire culture, 523–26, 534; legislation regarding, 521; Native Americans and, 516, 518; overview, 513–14; in prehistorical periods, 514–16, 515; ranchers and, 523–26, 534; retaliatory, 521; suppression of, 493, 517–18. See also prescribed burning
flameleaf sumac, 27, 338, 339, 480, 500
fleabane, 24, 142, 143
floodplains. See riparian areas
food. See diet and nutrition
forage growth, relationship with lightning strikes, 515, 516
forage inventories, 494, 495, 496
foraging behavior, 13–14
forbs, 24–27, 539, 540, 541, 552. See also *specific plants by scientific or common name*
*Forestiera pubescens*, 27, 400, 401, 556
Foster, J. H., 520–21

four-o'clock family. *See* Nyctaginaceae
fourwing saltbush, 27, *352,* 353
fragrant mimosa, 27, *374,* 375, 479
Froebel, Julius, 519
fuel, for prescribed burning, 503, *523–24,* 528–29
*Funastrum crispum,* 28, *432,* 433

*Gaillardia pinnatifida,* 24, *146,* 147
*Gaillardia pulchella,* 24, *148,* 149
*Gaura calcicola,* 26, *292,* 293
gayfeather, 24, *166,* 167, *510,* 557
Gentianaceae, 25. See also *Centaurium beyrichii*
geography, of Concho Valley, 3, 4, **5**
Geraniaceae: *Erodium cicutarium,* 25, *248,* 249; *Erodium texanum,* 25, *250,* 251
geranium family. *See* Geraniaceae
*Glandularia bipinnatifida,* 26, *328,* 329
*Glandularia pumila,* 26, *330,* 331
goathead, 27, *336,* 337
goats, 497, 540
goosefoot family. *See* Chenopodiaceae
gourd family. *See* Cucurbitaceae
grape family. *See* Vitaceae
grasses and grasslike plants, 23–24. See also *specific plants by scientific or common name*
grass establishment, 503–4
grass family. *See* Poaceae
grassland croton, 25, *208,* 209, 503
Gray, Andrew, 519–20
gray golden aster, 24, *162,* 163
grazing management principles, 494–95, *496,* 505–6, 522–23, *524,* 541. *See also* overgrazing
greenbriar, 28, *450,* 451, 556
greenbriar family. *See* Smilacaceae
green condalia, 27, *404,* 405, *500,* 556
green sprangletop, 23, *80,* 81
greenthread, 25, *184,* 185
*Grindelia nuda,* 24, *150,* 151
*Grindelia papposa,* 24, *152,* 153
ground cherry, 26, *318,* 319, 557

*Gutierrezia dracunculoides,* 24, *154,* 155, 557
*Gutierrezia sarothrae,* 14, 24, *156,* 157–59, *158,* 502
"guzzlers," 558
*Gymnosperma glutinosum,* 24, *160,* 161

habitat management: doves, 548–49; quail, 550–52, 554; songbirds, 556, 557–60; turkeys, 545; white-tailed deer, 538–41
hackberry tree, 497, *498, 500,* 556
hairy grama, 23, *52,* 53
hairy tridens, 23, *76,* 77
hairy tube tongue, 24, *118,* 119
Hall panicum, 23, *86,* 87
heath aster, 24, *138,* 139
*Hedeoma drummondii,* 25, *254,* 255, 502
hedging, 498–99
*Hedyotis nigricans,* 26, *310,* 311
Heller's plantain, 26, *304,* 305
herd management, of white-tailed deer, 537–38
*Heterotheca canescens,* 24, *162,* 163
*Hilaria belangeri,* 23, *78,* 79
hoary euphorbia, 25, *206,* 207
hogs. *See* feral hogs
holly family. *See* Berberidaceae
Homestead Act of 1862, 517
honey mesquite, 27, *376,* 377–79, *378,* 493, 519, 556
honeysuckle family. *See* Caprifoliaceae
hooded windmillgrass, 23, *62,* 63
horse crippler, 29, *462,* 463
horses, 517
Huisache daisy, 24, *128,* 129, 502
hummingbirds, *507,* 556, *562*
*Hymenopappus artemisiifolius,* 24, *164,* 165

*Ibervillea lindheimeri,* 28, *444,* 445
Inca doves, 549
Indian blanket, 24, *148,* 149
Indiangrass, *8*
Indian mallow, 26, *270,* 271, 557
Indians, use of fire by, 516, 518

insecticides, 560
introduced species, defined, 17
invasive plants, 14, *15*, 16–17
Invasive Species Policy (NRCS), 16
*Ipomoea cordatotriloba*, 28, *438*, 439
*Ipomoea lindheimeri*, 28, *440*, 441
irrigation systems, 560
ivy treebine, 28, *452*, 453, 556

javelina bush, 27, *402*, 403
jimsonweed, 3, 26, *316*, 317
Juglandaceae: *Carya illinoinensis* (*See* pecan); *Juglans microcarpa*, 27, *398*, 399, *480*, 504
*Juglans microcarpa*, 27, *398*, 399, *480*, 504
Jumano Indians, 493
*Juniperus ashei*. *See* Ashe juniper
*Juniperus pinchotii*. *See* redberry juniper
juvenile brush species, prescribed burning of, 528

Kickapoo Creek, 488, 489, 490
Kiowa Creek, 491
knotgrass, 23, *94*, 95, 504
knotweed leaf-flower, 25, *218*, 219, 557
Komarek, Ed, 513
Krameriaceae, 25. See also *Krameria lanceolata*
*Krameria lanceolata*, 25, *252*, 253

lacy cactus, 29, *466*, 467
lambsquarter, 557
Lamiaceae: *Hedeoma drummondii*, 25, *254*, 255, 502; *Marrubium vulgare*, 25, *256*, 257, 502; *Monarda citriodora*, 25, *258*, 259, 502; *Salvia farinacea*, 25, *260*, 261, 502; *Salvia reflexa*, 26, *262*, 263, 502; *Salvia texana*, 26, *264*, 265; *Scutellaria drummondii*, 26, *266*, 267
lance-leaf sage, 26, *262*, 263, 502
landscape features: browse lines, 497–98; grass establishment, 503–4; hedging, 498–99; historical accounts of, 485–93; mottes, 499–500; plants identification through smell, 502–3; reading, 497–504; resprouting, 500, *501;* riparian areas, 7, *484*, 504–6, *507–12*
laws and legal issues: baiting laws, 548; fires, 521; "law of the range," 522
lazy daisy, 24, *134*, 135
leatherweed croton, 25, *212*, 213, 503
legal issues. *See* laws and legal issues
legume family. *See* Fabaceae
lemon beebalm, 25, *258*, 259, 502
*Lepidium virginicum*, 25, *196*, 197, 557
*Leptochloa dubia*, 23, *80*, 81
*Liatris punctata*, 24, *166*, 167, *510*, 557
lightning strikes, relationship with forage growth, *515*, 516
Liliaceae: *Allium drummondii*, 23, *36*, 37, 503, 557; *Cooperia drummondii*, 23, *38*, 39
lily family. *See* Liliaceae
limestone gaura, 26, *292*, 293
Limestone Hill, 10
Lindheimer morning glory, 28, *440*, 441
Lindheimer's copperleaf, 25, *204*, 205
Lipan Creek, 489
little bluestem, 24, *98*, 99, *530*, 551
littleleaf sumac, 27, *340*, 341, *500*, 552
little walnut, 27, *398*, 399, *480*, 504
live oak: browse lines of, 497; classification of, 27; historical accounts of, 487–88, 489; profile of, *392*, 393–95, *394;* resprouting capabilities of, 515; in riparian areas, *508;* soil conditions for, 10; suppressed growth of, 519; toxicity of, 14
livestock: grazing management principles, 494–95, *496*, 505–6, 522–23, *524;* introduction to Concho Valley, 493
loafing cover, 552
Loasaceae, 26. See also *Mentzelia oligosperma*
*Lobelia cardinalis*, 25, *198*, 199, *507*

longhorn cattle, 490, 493
*Lonicera albiflora*, 27, *350*, 351, 556
lotebush, 27, *406*, 407, 552, 556
low menodora, 26, *288*, 289, 557
Low Stony Hill, 10
low wild mercury, 25, *214*, 215, 556–57
*Lupinus subcarnosus*, 25, *234*, 235, 557
*Lycium berlandieri*, 28, *418*, 419, 556
*Lygodesmia texana*, 24, *168*, 169
lyreleaf parthenium, 24, *174*, 175

*Machaeranthera pinnatifida*, 24, *170*, 171
madder family. *See* Rubiaceae
Major Land Resource Areas (MLRAs), 8–10, **9**
mallow family. *See* Malvaceae
malta star-thistle, 16
Malvaceae: *Abutilon fruticosum*, 26, *270*, 271, 557; *Callirhoe involucrata*, 26, *272*, 273, 557; *Rhynchosida physocalyx*, 26, *274*, 275, 557; *Sida abutifolia*, 26, *276*, 277; *Sphaeralcea angustifolia*, 26, *278*, 279; *Sphaeralcea coccinea*, 26, *280*, 281
*Marrubium vulgare*, 25, *256*, 257, 502
mast, as nutritional source, 539
Maximilian sunflower, *18*
mealycup sage, 25, *260*, 261, 502
*Medicago minima*, 25, *236*, 237
*Melampodium leucanthum*, 24, *172*, 173
*Melica nitens*, 23, *82*, 83
Mendoza, Juan Domínguez de, 485, 487–88
Menispermaceae, 28. *See also Cocculus carolinus*
*Menodora heterophylla*, 26, *288*, 289, 557
*Mentzelia oligosperma*, 26, *268*, 269
mesquite grass, 488, 489
Mexican sagewort, 24, *136*, 137, 502
Middle Concho River, 4, **5**, 487, 489, 491, 492, *511*
milkweed family. *See* Asclepiadaceae

milkwort family. *See* Polygalaceae
*Mimosa aculeaticarpa* var. *biuncifera*, 27, *372*, 373, 479
*Mimosa borealis*, 27, *374*, 375, 479
*Mimosa nuttallii*, 25, *238*, 239
mint family. *See* Lamiaceae
minty-smelling plants, 502
mistletoe, 28, *428*, 429, 556
mistletoe family. *See* Viscaceae
MLRAs (Major Land Resource Areas), 8–10, **9**
Mohr shin oak, 27, *382*, 383–85, *384*
monarch butterfly, *564*
*Monarda citriodora*, 25, *258*, 259, 502
moonseed family. *See* Menispermaceae
Mormon tea family. *See* Ephedraceae
morning glory family. *See* Convolvulaceae
*Morus rubra*, 556
mottes, 499–500
mountain pink, 25, *246*, 247
mourning doves, 548, *549*
mulberry, *508*, 556
mustard family. *See* Brassicaceae

narrowleaf globemallow, 26, *278*, 279
narrowleaf spiderling, 26, *284*, 285
*Nassella leucotricha*, 23, *84*, 85, 488
Native Americans, use of fire by, 516, 518
native species, defined, 17
Natural Resources Conservation Service (NRCS), 10, 15, 16
neighboring states, noxious plants in, 16
Nelle, Steve, 504
nests: of quail, 551–52; of songbirds, 557; of turkeys, 545–46
netleaf hackberry, 28, *422*, 423
nightshade family. *See* Solanaceae
Noelke soils, 10, **11**, *12*
*Nolina texana*, 29, *456*, 457
nonnative species, defined, 17
North Concho River, 4, **5**
noseburn, 25, *224*, 225, 557
noxious plants, 14–17, *15–16*

NRCS (Natural Resources Conservation Service), 10, 15, 16
nutrition. *See* diet and nutrition
Nuttall peavine, 25, *226*, 227
Nyctaginaceae: *Acleisanthes longiflora*, 26, *282*, 283; *Boerhavia linearifolia*, 26, *284*, 285; *Nyctaginia capitata*, 26, *286*, 287, 502
*Nyctaginia capitata*, 26, *286*, 287, 502

oak family. *See* Fagaceae
*Oenothera speciosa*, 26, *294*, 295
old man's beard, 28, *448*, 449
Oleaceae: *Forestiera pubescens*, 27, *400*, 401, 556; *Menodora heterophylla*, 26, *288*, 289, 557
olive family. *See* Oleaceae
Olmsted, Frederick, 519
Onagraceae: *Calylophus hartwegii*, 26, *290*, 291; *Gaura calcicola*, 26, *292*, 293; *Oenothera speciosa*, 26, *294*, 295
one-seed croton, 25, *210*, 211, 503
*Opuntia engelmannii*. *See* prickly pear
*Opuntia leptocaulis*, 29, *472*, 473, 528, 556
orange zexmenia, 25, *190*, 191, 478
Ormsby, Waterman L., 487, 491–92
overgrazing, 520, 540
Oxalidaceae: *Oxalis drummondii*, 26, *296*, 297; *Oxalis stricta*, 26, *298*, 299, 556
*Oxalis drummondii*, 26, *296*, 297
*Oxalis stricta*, 26, *298*, 299, 556

packrat middens, 485
*Panicum hallii*, 23, *86*, 87
*Panicum obtusum*, 23, *88*, 89, 551
*Panicum virgatum*, 23, *90*, 91, 504, 509–10
Papaveraceae, 26. See also *Argemone aurantiaca*
Parker, Amos, 518
parsley family. *See* Apiaceae
*Parthenium confertum*, 24, *174*, 175
*Parthenocissus quinquefolia*, 28, *454*, 455, 556
*Pascopyrum smithii*, 23, *92*, 93

*Paspalum distichum*, 23, *94*, 95, 504
pecan: classification of, 27; historical accounts of, 487–88, 489; leaves of, *480*; profile of, *396*, 397; in riparian areas, 504, *508*; as songbird nutrition, 556
Pecan Creek, 4, 489
penstemon, 26, *314*, 315
*Penstemon cobaea*, 26, *314*, 315
pepperweed, 25, *196*, 197, 557
perennial grass communities, 528
persimmon family. *See* Ebenaceae
pests, defined, 17
*Phacelia congesta*, 7
*Phoradendron tomentosum*, 28, *428*, 429, 556
photographic guidelines, 12
*Phyla nodiflora*, 27, *332*, 333
*Phyllanthus polygonoides*, 25, *218*, 219, 557
*Physalis cinerascens*, 26, *318*, 319, 557
Phytolaccaceae, 26. See also *Rivina humilis*
pigeonberry, 26, *302*, 303, 557
pigweed, 556
pincushion daisy, 24, *146*, 147
pink verbena, 26, *330*, 331
pipevine family. *See* Aristolochiaceae
plains bristlegrass, 24, *100*, 101
plains lovegrass, 23, *72*, 73
plains zinnia, 25, *192*, 193
Plantaginaceae: *Plantago helleri*, 26, *304*, 305; *Plantago rhodosperma*, 26, *306*, 307
*Plantago helleri*, 26, *304*, 305
*Plantago rhodosperma*, 26, *306*, 307
plantain family. *See* Plantaginaceae
plants: classification of, 12–13; deer food plants, *542–44*; ethnobotany of, 17; for hummingbirds and butterflies, *562–63*; identification through smell, 502–3; noxious and invasive, 14–17, *15–16*; quail food plants, *553*; riparian, *512*; toxic/poisonous, 13–14; turkey food plants, *547*. See also *specific plants and plant families*

*Pleuraphis mutica*, 23, *96*, 97, 551
Poaceae: *Andropogon gerardii*, 23, *40*, 41; *Andropogon glomeratus*, 23, *42*, 43, 504; *Aristida purpurea* var. *longiseta*, 23, *44*, 45; *Aristida purpurea* var. *wrightii*, 23, *46*, 47; *Bothriochloa laguroides*, 23, *48*, 49; *Bouteloua curtipendula*, 23, *50*, 51, 551; *Bouteloua hirsuta*, 23, *52*, 53; *Bouteloua rigidiseta*, 23, *54*, 55; *Bouteloua trifida*, 23, *56*, 57; *Bromus catharticus*, 23, *58*, 59; *Buchloe dactyloides*, 23, *60*, 61; *Chloris cucullata*, 23, *62*, 63; *Chloris verticillata*, 23, *64*, 65; *Digitaria cognata*, 23, *66*, 67; *Elymus canadensis*, 23, *68*, 69; *Elymus longifolius*, 23, *70*, 71; *Eragrostis intermedia*, 23, *72*, 73; *Eriochloa sericea*, 23, *74*, 75; *Erioneuron pilosum*, 23, *76*, 77; *Hilaria belangeri*, 23, *78*, 79; *Leptochloa dubia*, 23, *80*, 81; *Melica nitens*, 23, *82*, 83; *Nassella leucotricha*, 23, *84*, 85, 488; *Panicum hallii*, 23, *86*, 87; *Panicum obtusum*, 23, *88*, 89, 551; *Panicum virgatum*, 23, *90*, 91, 504, 509–10; *Pascopyrum smithii*, 23, *92*, 93; *Paspalum distichum*, 23, *94*, 95, 504; *Pleuraphis mutica*, 23, *96*, 97, 551; *Schizachyrium scoparium*, 24, *98*, 99, *530*, 551; *Setaria leucopila*, 24, *100*, 101; *Setaria reverchonii*, 24, *102*, 103; *Sorghastrum nutans*, 24, *104*, 105; *Sporobolus compositus*, 24, *106*, 107, *529*; *Sporobolus cryptandrus*, 24, *108*, 109; *Tridens albescens*, 24, *110*, 111; *Tridens muticus* var. *muticus*, 24, *112*, 113; *Tripsacum dactyloides*, 24, *114*, 115, 504, *510*
poison ivy, 28, *430*, 431, 556
poisonous plants, 13–14
pokeweed family. *See* Phytolaccaceae
*Polygala alba*, 26, *308*, 309
Polygalaceae, 26. *See also Polygala alba*

poppy family. *See* Papaveraceae
prairie dogs, 488–89, 490
prairie verbena, 26, *328*, 329
precipitation, in Concho Valley, 4, 6–7, *8*
predator control, 546
prescribed burning, 503, 518, 524–29, *528–34*, 540, 559–60
prickly pear: classification of, 29; profile of, *468*, 469–71, *470*; quail's use of, 552; as songbird nutrition, 556; suppression through prescribed burning, 527–28, *531–32*
*Prosopis glandulosa*, 27, *376*, 377–79, *378*, 493, 519, 556
purple dalea, 25, *228*, 229
purple eryngo, 24, *120*, 121, 557

quail, 550–54, 551, *553*
*Quercus macrocarpa*, 27, *380*, 381, *508*
*Quercus mohriana*, 27, *382*, 383–85, *384*
*Quercus pungens* var. *vaseyana*, 27, *386*, 387, 389
*Quercus sinuata* var. *breviloba*, 14, 27, *390*, 391
*Quercus virginiana*. *See* live oak

rabbit tobacco, 24, *144*, 145
rain lily, 23, *38*, 39
ranchers: fire culture of, 523–26, *534*; grazing management principles used by, 522–23
Ranunculaceae, 28. *See also Clematis drummondii*
raptors, 554
ratany family. *See* Krameriaceae
rat-ear coldenia, 25, *194*, 195, 557
*Ratibida columnifera*, 24, *176*, 177, *564*
rattlesnakes, 488, 489
reading landscapes, 497–504
redberry juniper: classification of, 27; management of, 495; profile of, *358*, 359–61, *360*; resprouting capabilities of, 495, 500, *501*, 527; smell emitted by, 502
red grama, 23, *56*, 57

red mulberry, 508
red-seed plantain, 26, *306*, 307
red threeawn, 23, *44*, 45
regulations. *See* laws and legal issues
reproduction, of songbirds, 561
rescue grass, 23, *58*, 59
resprouting, 500, *501*
rest and rotation system, in grazing management, 494
Reverchon bristlegrass, 24, *102*, 103
Rhamnaceae: *Condalia ericoides*, 27, *402*, 403; *Condalia hookeri*, 27, *404*, 405, 500, 556; *Ziziphus obtusifolia*, 27, *406*, 407, 552, 556
*Rhus copallinum*, 27, *338*, 339, 480, 500
*Rhus microphylla*, 27, *340*, 341, 500, 552
*Rhus trilobata*, 27, *342*, 343, 500, 502
*Rhus virens*, 27, *344*, 345
*Rhynchosia senna*, 25, *240*, 241
*Rhynchosida physocalyx*, 26, *274*, 275, 557
Rio Grande turkeys, wildlife management for, 544–47, 545
riparian areas, 7, 484, 504–6, 507–12
riparian sponge, 504–5
rivers, of Concho Valley, **486**
*Rivina humilis*, 26, *302*, 303, 557
rock daisy, 24, *172*, 173
Roemer's acacia, 27, *368*, 369, 479
Rolling Limestone Prairie, 8, **9**
Rolling Red Plains, 8, **9**
rotation system, in grazing management, 494
Rubiaceae: *Cephalanthus occidentalis*, 28, *408*, 409, 504; *Hedyotis nigricans*, 26, *310*, 311
ruellia, 24, *116*, 117, 557
*Ruellia metziae*, 24, *116*, 117, 557
Rutaceae: *Thamnosma texana*, 26, *312*, 313, 502; *Zanthoxylum hirsutum*, 28, *410*, 411, 500, 502, 556

sacahuista, 29, *456*, 457
sage-smelling plants, 502
Salicaceae, 28. *See also Salix nigra*
*Salix nigra*, 28, *412*, 413, 504
*Salsola tragus*, 25, *200*, 201–3, *202*
salt cedar, 28, *420*, 421
salt cedar family. *See* Tamaricaceae
*Salvia farinacea*, 25, *260*, 261, 502
*Salvia reflexa*, 26, *262*, 263, 502
*Salvia texana*, 26, *264*, 265
sand dropseed, 24, *108*, 109
Sapindaceae, 28. *See also Sapindus saponaria* var. *drummondii*
*Sapindus saponaria* var. *drummondii*, 28, *414*, 415, 480, 502, 508
Sapotaceae, 28. *See also Sideroxylon lanuginosum*
sawgrass, 510
sawleaf daisy, 24, *152*, 153
scaled quail, 550, *551*
scarlet globemallow, 26, *280*, 281
scarlet musk flower, 26, *286*, 287, 502
*Schizachyrium scoparium*, 24, *98*, 99, 530, 551
Scrophulariaceae, 26. *See also Penstemon cobaea*
*Scutellaria drummondii*, 26, *266*, 267
Sedge family. *See* Cyperaceae
semiarid climates, 4
*Senecio douglasii*, 24, *178*, 179
*Senna roemeriana*, 25, *242*, 243
sensitive briar, 25, *238*, 239
*Setaria leucopila*, 24, *100*, 101
*Setaria reverchonii*, 24, *102*, 103
sheep, 490, 493, 523, 540
shin oak, 500, *501*, 515, *529*
showy evening primrose, 26, *294*, 295
shrubs and trees, 27–28. *See also specific plants by scientific or common name*
*Sida abutifolia*, 26, *276*, 277
sideoats grama, 23, *50*, 51, 551
*Sideroxylon lanuginosum*, 28, *416*, 417, 556
signal grass, 557
silver bluestem, 23, *48*, 49
silverleaf nightshade, 26, *322*, 323
*Simsia calva*, 24, *180*, 181, 478
*Siphonoglossa pilosella*, 24, *118*, 119
skullcap, 26, *266*, 267
skunkbush sumac, 27, *342*, 343, 500, 502

slender vervain, 27, *334*, 335
slim tridens, 24, *112*, 113
smell, plant identification through, 502–3
Smilacaceae, 28. See also *Smilax bona-nox*
*Smilax bona-nox*, 28, *450*, 451, 556
snags, 560
snow-on-the-mountain, 25, *216*, 217
soapberry family. *See* Sapindaceae
soils, in Concho Valley, 10, **11**, *12*, 504
Solanaceae: *Datura stramonium*, 3, 26, *316*, 317; *Lycium berlandieri*, 28, *418*, 419, 556; *Physalis cinerascens*, 26, *318*, 319, 557; *Solanum dimidiatum*, 3, 26, *320*, 321; *Solanum elaeagnifolium*, 26, *322*, 323; *Solanum rostratum*, 26, *324*, 325; *Solanum triquetrum*, 26, *326*, 327
*Solanum dimidiatum*, 3, 26, *320*, 321
*Solanum elaeagnifolium*, 26, *322*, 323
*Solanum rostratum*, 26, *324*, 325
*Solanum triquetrum*, 26, *326*, 327
*Solidago canadensis*, 25, *182*, 183, 557
songbirds, 556–61; cover types used by, 557–58; diet of, 556–57, *562*; habitat management for, 556, 557–60; nests of, 557; reproduction of, 561; water for, 558
*Sorghastrum nutans*, 24, *104*, 105
sour-smelling plants, 502
South Concho River, 4, **5**, 489, 490, *508–9*
Southern Desertic Basins, Plains, and Mountains, 8, **9**
Southern High Plains, 8, **9**
*Sphaeralcea angustifolia*, 26, *278*, 279
*Sphaeralcea coccinea*, 26, *280*, 281
spike rush, 23, *34*, 35, 504
spiny bumelia, 28, *416*, 417, 556
*Sporobolus compositus*, 24, *106*, 107, 529
*Sporobolus cryptandrus*, 24, *108*, 109
spreading sida, 26, *276*, 277

Spring Creek, 4, **5**, 489, 490, *509–10*
spurge family. *See* Euphorbiaceae
squirreltail, 23, *70*, 71
stickleaf, 26, *268*, 269
stickleaf family. *See* Loasaceae
sticky selloa, 24, *160*, 161
*Stillingia texana*, 25, *220*, 221
*Stillingia treculiana*, 25, *222*, 223
stocking rate, 494
strawberry cactus, 29, *464*, 465
streams, of Concho Valley, **486**
sumac family. *See* Anacardiaceae
sunflower family. *See* Asteraceae
sweet-smelling plants, 502
switchgrass, 23, *90*, 91, 504, *509–10*

tall dropseed, 24, *106*, 107, 529
tall goldenrod, 25, *182*, 183, 557
Talpa soils, 10, **11**
Tamaricaceae, 28. See also *Tamarix gallica*
*Tamarix gallica*, 28, *420*, 421
Tarrant soils, 10, **11**, *12*
tasajillo, 29, *472*, 473, 528, 556
taxonomy, for plants, 13
Taylor, Charles "Butch," 493, 513
Texas bindweed, 28, *436*, 437
Texas bluebonnet, 25, *234*, 235, 557
Texas Commission on Environmental Quality (TCEQ), 525
Texas cupgrass, 23, *74*, 75
Texas filaree, 25, *250*, 251
Texas frogfruit, 27, *332*, 333
Texas grama, 23, *54*, 55
Texas nightshade, 26, *326*, 327
Texas persimmon, 27, *362*, 363, 500
Texas salvia, 26, *264*, 265
Texas skeleton plant, 24, *168*, 169
Texas snoutbean, 25, *240*, 241
Texas stillingia, 25, *220*, 221
Texas wintergrass, 23, *84*, 85, 488
*Thamnosma texana*, 26, *312*, 313, 502
*Thelesperma filifolium*, 25, *184*, 185
threadleaf groundsel, 24, *178*, 179
three-flower melic, 23, *82*, 83
*Thymophylla pentachaeta*, 25, *186*, 187
tickle tongue, 28, *410*, 411, 500, 502, 556

*Tiquilia canescens*, 25, *194*, 195, 557
tobosagrass, 23, *96*, 97, 551
*Toxicodendron radicans*, 28, *430*, 431, 556
toxic plants, 13–14
*Tragia ramosa*, 25, *224*, 225, 557
trailing ratany, 25, *252*, 253
trecul stillingia, 25, *222*, 223
trees and shrubs, 27–28. See also *specific plants by scientific or common name*
*Tribulus terrestris*, 27, *336*, 337
*Tridens albescens*, 24, *110*, 111
*Tridens muticus* var. *muticus*, 24, *112*, 113
*Tripsacum dactyloides*, 24, *114*, 115, 504, *510*
tumbleweed, 25, *200*, 201–3, *202*
tumble windmillgrass, 23, *64*, 65
turkeys, wildlife management for, 544–47, *545*
Twin Buttes watershed, 4, **5**
twin-leaf senna, 25, *242*, 243

Ulmaceae: *Celtis laevigata* var. *reticulata*, 28, *422*, 423; *Ulmus americana*, 28, *424*, 425, *508*
*Ulmus americana*, 28, *424*, 425, *508*
United States Department of Agriculture (USDA), 10, 15
upright prairie coneflower, 24, *176*, 177, *564*
*Urochloa platyphylla*, 557

Vasey shin oak, 27, *386*, 387, 389
vegetation: of Concho Valley, 485–93; in riparian areas, 504–5
velvet bundleflower, 25, *232*, 233
Verbenaceae: *Aloysia gratissima*, 28, *426*, 427, 502; *Glandularia bipinnatifida*, 26, *328*, 329; *Glandularia pumila*, 26, *330*, 331; *Phyla nodiflora*, 27, *332*, 333; *Verbena halei*, 27, *334*, 335
verbena family. See Verbenaceae
*Verbena halei*, 27, *334*, 335
*Verbesina encelioides*, 25, *188*, 189, 557
vervain family. See Verbenaceae

*Vicia ludoviciana*, 25, *244*, 245
vine-mesquite, 23, *88*, 89, 551
vines, 28. See also *specific plants by scientific or common name*
Virginia creeper, 28, *454*, 455, 556
Viscaceae, 28. See also *Phoradendron tomentosum*
Vitaceae: *Cissus incisa*, 28, *452*, 453, 556; *Parthenocissus quinquefolia*, 28, *454*, 455, 556

walnut family. See Juglandaceae
water: for deer, 541; for doves, 549; for quail, 552; for songbirds, 558; storage tanks for, 523; for turkeys, 546
water birds, 554
wavy-leaf milkvine, 28, *432*, 433
Webb, Walter Prescott, 519
*Wedelia texana*, 25, *190*, 191, 478
weeds, noxious, 17
Western Edwards Plateau, 8, **9**, 515, 519, 522
western horsenettle, 3, 26, *320*, 321
western primrose, 26, *290*, 291
western ragweed, 24, *132*, 133, 477, 503
western scrub jay, *561*
western soapberry, 28, *414*, 415, 480, 502, *508*
western wheatgrass, 23, *92*, 93
whitebrush, 28, *426*, 427, 502
white honeysuckle, 27, *350*, 351, 556
white milkwort, 26, *308*, 309
white pricklypoppy, 26, *300*, 301, 557
white shin oak, 14, 27, *390*, 391
white-tailed deer, 537–42; browse lines created by, 497–98; culling practices and, 538; diet of, 538–40, 541–42, *542*–44; examples of, *536*, *539*; habitat management for, 538–41; herd management of, 537–38; riparian area damage by, 506
white tridens, 24, *110*, 111
white-winged doves, 548, 549
wildfires, 528–29

wild hogs. *See* feral hogs
wildlife management, 537–63; doves, 548–50, *549*; quail, 550–54, 551, *553*; songbirds, *555*, 556–61, 562–63; turkeys, 544–47, *545*; white-tailed deer, 537–42, *539*, 542–44
wild onion, 23, *36*, 37, 503, 557
wild petunia family. *See* Acanthaceae
willow baccharis, 27, *346*, 347
willow family. *See* Salicaceae
windmills, 523
wine cup, 26, *272*, 273, 557
wolfberry, 28, *418*, 419, 556
wolf-proof fences, 523
woodpeckers, 556
woodsorrel, 26, *298*, 299, 556

woodsorrel family. *See* Oxalidaceae
Wright threeawn, 23, *46*, 47

XIT ranch, 517–18

yellow Indiangrass, 24, *104*, 105
yellow star-thistle, 16
*Yucca constricta,* 29, *458*, 459–61, *460*

*Zanthoxylum hirsutum,* 28, *410*, 411, 500, 502, 556
*Zinnia grandiflora,* 25, *192*, 193
*Ziziphus obtusifolia,* 27, *406*, 407, 552, 556
Z-L Ranch, 521
Zygophyllaceae, 27. *See* also *Tribulus terrestris*